Physics of M and Yang

– to make Einstein's dreams
come true

Kwang-gyu Yoon

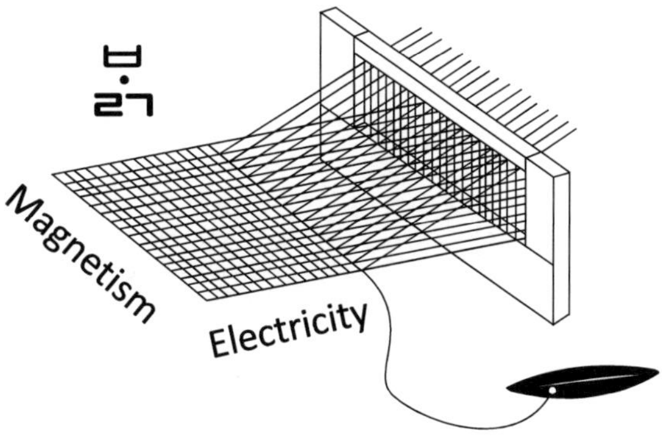

About a language

Korean language, which is the writer's mother tongue as well, is one of the oldest in the world but still young a language. To interpret the mentioned Korean language's characteristic, it began as a written and artificial one. The language seems to have become a verbal one quite recently and owes its verbalization to King Sejong the great, who had crafted the phonetic values of Korean alphabet. His intention of inventing Korean alphabet was primarily to define the pronunciations of Chinese letters. Chinese character had been the linguistic means for his men and people to communicate by. It was, and still is, a great treasure box in which cultural, philosophical, historical jewels are conserved but had a fatal shortcoming for communication. There is no Chinese letter that has put any clue of how to pronounce it on itself. His people from countries were unable to make understood what they wished to speak in the way of arraying Chinese letters, because people's vocalizations were so seriously different from one another's that their pronunciations of Chinese letters failed to make sense to the government officials as well as his majesty. He had a deep compassion with his pitiful people. It seemed obvious to him that there should have been a suit of standardized phonetic symbols for Chinese letters, just as modern China has adopted Roman alphabet to denote their languages, whose sentences are composed of Chinese letters. There had already been alphabetic letters that would wear the pronunciations, whose looks were not so different from the present Korean alphabet. What mattered was how to find the original sound of the letters. Sejong the great did it anyway and published it. However, his alphabet was not popular after his death.

The vocabulary of Korean language, excluding words adopted from European languages recently, mostly from English, is composed of 30 percents of so called indigenous Korean words and 70 percents of Chinese letters. What is meant by indigenous Korean is not that Korean people had been speaking indigenous Korean language in a way and imported words of Chinese character into their own language. It was mentioned that the King's alphabet was invented to denote how to verbalize Chinese letter correctly. Indigenous Korean words were denotations of meanings of Chinese letters. Phonetic symbols are used to denote a pronunciation of a word. The King's alphabet was mobilized once again to write down what a Chinese letter means. In other words, so called indigenous Korean words had existed before the Chinese character was created. The alphabet was a suit of phonetic symbols as well as semantic symbols for Chinese letters.

Around the collapse of his kingdom about three hundreds years later, the whole territory fell in a chaotic situation in part for its self contradictions and more for western powers' invasions. Meanwhile, the king's people scattered and gathered repeatedly. Some portion of them ended up in Korean peninsula and formed an ethnic community bearing the spirit and the culture of the nation Sejong the great hoped to model. Korean language had finally settled on Korean people's tongues through the times of turbulence.

Korean language is a creation craftily devised in the principle of yin and yang(M and Yang, according to his majesty's publication). The discovery of this fact itself is quite an advance and an enhance -ment of motivation to trace truths. Otherwise the publication of this paper would be impossible.

■ King Sejong the great (1391-1450)
■ Lee Dyou(李道).
The forth King of Dyou-Shyun (Cho-suhn, 1392-1910)

He has many titles of 'the first in the world'. The purpose of his alphabet, now Korean alphabet, was very democratic, in order to have people read. A lot of Buddhism, Confucianism scriptures, literary works, and other materials were interpreted in his alphabet, which lead to ampleness of Korean language. He was an inventor and the manager of inventors. The most representative invention is the hydro-operation automatic clock. It actually conducted its duty to notify the exact standard time for about two hundred years until Japanese invasion(1592-1599). His government operated to the clock. He ordered public opinion surveys for new policies, presumably never been before in the world. Even slaves were given a maternity leave together with the husband by law. There was officially no noble class in his kingdom.

He was a sincere Buddhist and ardent Confucian scholar, as well as a greatest scientist in human history. A vase of flower to have blossomed earlier than usual was set on his table. He asked a courtier about the flower. It was from a green house - there was the green house in the court, according to the record. He ordered doing away with the flower and said, "Seeing a flower in a wrong season is very against principles in nature." It is said the green house was usually for medical herb raising.

Physics of M and Yang

- to make Einstein's dreams come true

윤 광 규
Kwang-gyu Yoon

Contents

Preface

The dreams made true at this paper are the completion of unified theory of field and beating Niels Bohr in the battle of quantum mechanics. Unfortunately, his theory of relativity will be disposed of too. However, I like Niels Bohr more than Albert Einstein. There is one kind of force in the universe, which is generated with M and Yang. So the first dream is already achieved. This does not mean this paper will not talk about what it is. It will be quite an excitement. Bohr is the very physicist that ignited my will to fight and the spirit of second to none. M and Yang has been something like air culturally and philosophically ever since I was born in Korean society. His mention of Tai-gg(Tai-chi), which is composed of M and Yang, made me grit my teeth. It felt like my patent right was infringed. Rumor has it that he branded his knighthood sword with Tai-gg pattern. And I did not like his complementarity concept ad-lib. Simply, M and Yang is not such a thing he believed to be. It was not the first encountering him. The first one happened in a physics class when I was a high school student. My teacher introduced atomic models suggested by the early physicists. Niels Bohr's was one of them. I saw a critical contradiction in his model but the teacher did not mention that. The curiosity was not dissolved even while I learned electronics at college. Rather weirdly, other eminent scientists, including Einstein, had praised him and his idea of the model. His remark of the Eastern philosophy, M and Yang, to complement his quantum mechanics (It sounded like a misuse to me) was the moment of my resolution thinking, 'I've gotta stop these guys!'.

A specific study of M and Yang has not ever been fulfilled modern times. It has been regarded as a metaphysics or a philosophy. It felt like I was lost even though I had declared I would beat the westerners. There was no book, no theories, no teacher, talking about M and Yang concretely. It was so desperate and helpless. In the meantime I ran into an old writing about Korean alphabet. It was a part of the manual book Hyerye which accounted for how the alphabet should be used, its background philosophy, and so on. It said the alphabet was created on the basis of M and Yang principle. It seems Korean language had been a written one till the time. Thankfully, king Sejong the great, who created the alphabet, left such a lot of works written in Korean language by himself or by order. It is such a blessing to all that a descendant whose mother tongue is Korean language can have a opportunity, through the alphabet, words and expressions written in it, to research how M and Yang actually works. It gives students of M and Yang more relief since Korean language has barely survived. Homage to King Sejong the great and to all who put their everything to keep and inherit Korean language.

Though two 20th century representative physicists have been

mentioned for criticism, it does not mean the classical mechanics, such as Newton's and Kepler's, make good sense in light of M and Yang. It was just because two big figures of modern science were mentioned first in order. To belt out about the classic, first of all, it's not that Earth is flat, but that it ought to be flat. Upon the Earth we believe is so now, gravity and light are unable to generate; thereby so are matter and even antimatter; Coulomb's, Faraday's, Ampere's, Gauss' law, and Fleming's left hand law can not stand; Newton mechanics, theories of relativity, and quantum mechanics are impossible to justify; all the chemical actions and reactions would not take place. In a nutshell, all the things we experience could not occur in a model of round Earth. In it even our beings are impossible. Now it is time to get back to ask ancient wise men.

January 2020
At city Daegu, Korea

Acknowledgements

Unlike the modern trend of evolusionism, the faith says that languages are not products of nature. Languages' being itself testifies that evolusionism of whatever is even unscientific. They are a product of hyper-reasoning, not what can be assembled into life by a long time flowing. There are religions behind all the existing languages. Even the beginning of what science and mathematics are is from religions. To express accurately, religion and science were not distictive areas at the beginning. The principle of M and Yang is a good example. It explains matter as if it is metaphysical or absractions as if they have a concrete body. Thereby self-smart but fools tended to denounce the realistic being of M and Yang. History is full of absurdities. Korean alphabet was disdained even by seekers of M and Yang priciple, as it is an almost perfect relflection of M and Yang principle. It went without saying that Korean language expressed in the alphabet was disrespected and mistreated. The most terrible threat to Korean language was imperialists' supremacy of their own cultures. No matter what it might be, if it was not from Europe, it was seen as an inferior. During the colonial era by Japanese, Korean language was once banned from speaking in the public area. It was worth extermination to their proud eyes. But God helps. Korean language and the alphabet survived. Writing this paper, the writer was so moved with the language's being itself. No doubt it is so big a contribution to modern civilization.

Great homage to Sejong the great.

Heartfelt gratitude to all those who shed blood, sweat, and tears to keep the language and bequeath it to the next generation.

1

Light is not a particle

In spite of so much sweat scientists shed, the pursuits to truths were often misguided. The most notable one is Albert Einstein's light quantum experiment. Proving light is not a particle is meaningless in the world of M and Yang because there is no such thing and, in other words, what we see is not what it is really. But Judging by waht we see is also what we human can not help when trying to make an inference from the observation. The point of what was wrong with the light quantum theory is rather the experiment itself.

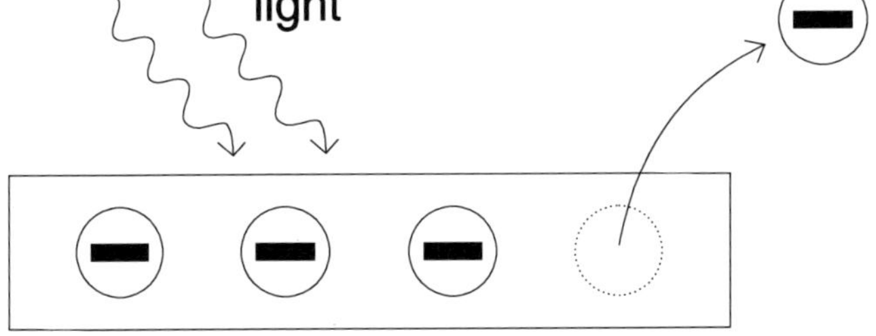

figure 1.1 This photoelectron experiment is a trick, though not intended by the experimenter

Light is what the experimenter wants to know in the experiment and an electron is the means. Einstein shed different sorts of light on the metal to see what happens. That is to say, he impacted the means with the object, intending to see what it is. Simply, that's wrong. Stimulating the object with the means and watching how the object reacts is reasonable and logical. He did it in the opposite. Therefore the conclusion should be invalidated, no matter what outcome the experiment produced. The result was that even a small amount of short wavelength light drove electrons to move, while long wavelength light failed it, however much it was shed on the metal bar. Such a wonder it can be a firm evidence that light be a particle. Secondly, the expression that light is a particle as well as a wave simultaneously is a semantic nonsense. You can be a father as well as a son but can not be male and female at the same time - Please, say not there are bisexuals. His theory of relativity has also been built on the wrong basis of light's duality. So what he discovered just proves that light is not a particle.

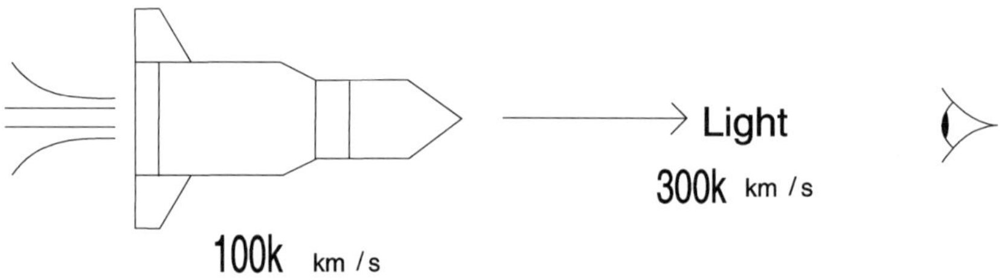

figure 1.2 no wonder the velocity of light is constant

It is said the velocity of light is constant regardless of its surroundings. The light shot from a rocket flying very fast is a typical example. It is said the velocity of the rocket does not add to that of light. The same phenomenon is observed on the light shot backward. It is not subtracted. And everyone wonders about that and says it is light's mysterious properties. Look, if the light were a definite particle, for example a bullet, one's wonder would be so natural. However, let's not look on light as a particle but just a definite wave this time. Then what we see? The constancy of velocity of light is way too natural. Why don't we take a sound to depart vehicles as an example?

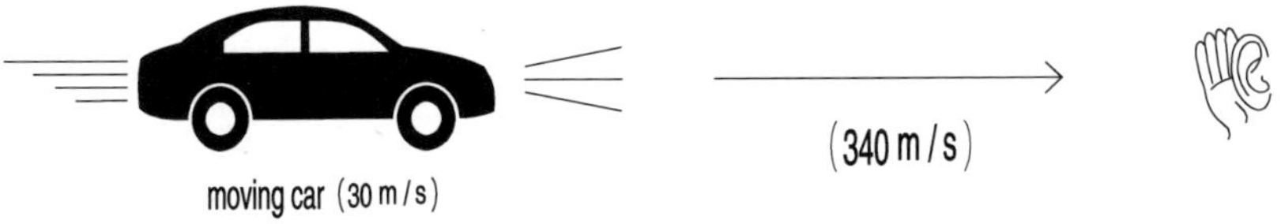

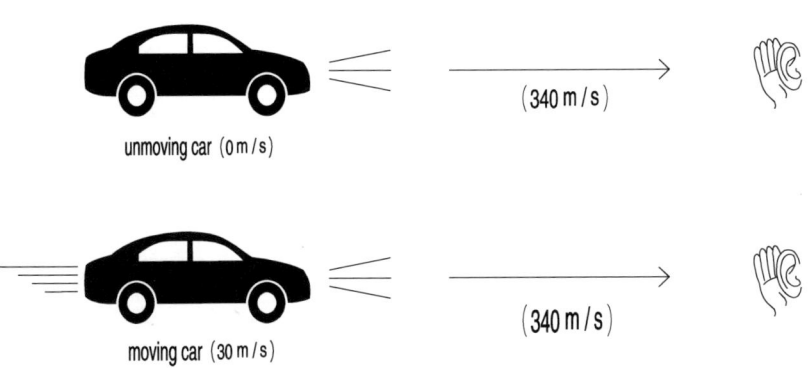

figure 1.3 the rash conclusion that light is also a particle makes constancy of light's velocity a wonder.

The velocities of the sounds are identical, whether it is from an unmoving car or from a moving car. Isn't this common sense? Light is a wave just like other ones such as sound, waves in water, and electromagnetic wave. Einstein's special relativity theory is well known for the hypothesis that time flows very slow in a very fast moving system like the light clock in a fast running train. But the thought experiment is also an reflection of the wrong idea that light is a particle. The properties of wave, diffraction and interference, are not considered in that experiment. The logical conclusion of his special and general relativities should be they are nothing but evidences denying light's duality. Einstein had gone counter ways. He employed the unproved conclusion(duality of light) as a basis and called the phenomenon(relativity theory) its conclusion.

There has been an inevitable reason for physicists' wish that light may probably be a particle. In terms of how many evidences each argument scores, light's waveness outdoes grainness. The only one supporting light being a particle is that light travels an empty space, so called vacuum. As believed commonly, light arrives on Earth through the outer space between the sun and Earth. The question was how a wave can come through a space of no medium for it. Thus, light should have been a particle. But that, all other things say light is a wave.

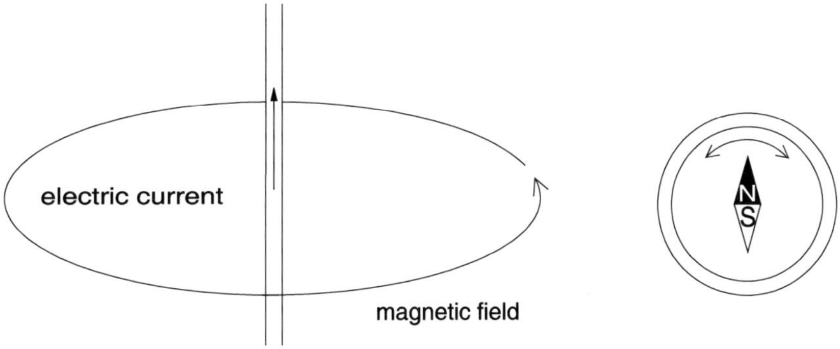

figure 1.4 science looks like an objective study but is quite similar to a religion. You don't discover something but do believe or not.

The right experiment to see what light is like should be something like the depict(figure 1.4). An electric current is a flowing of electrons, intuitively observed. Those electrons' motions induce electromagnetic wave in the space between the wire and the compass and cause the needle to shake. All agree light is also a sort of electromagnetic wave. That takes place in a vacuum too. Now let me suggest two options as the conclusion of this experiment.

Op 1. Light crossed the empty space. Therefore light is a particle.

Op 2. Light is a wave and crossed the space.
No wave is able to advance in a space without medium for it Therefore there is something in the space between the wire and the compass.

There had been an attempt to find a medium for light. Michelson and Morley's experiment was one of them. It was not found nor extracted, so called ether. Einstein adopted an idea that space can be warped and bent, however, nothingness of space has been taken for granted.

First of all, the question should be why we fail to find the medium of light, if any there. The easiest inference is a naming failure. Wave is a series of vibrations of medium. We see the medium but where is the wave? Do I see a wave really or it is our notion recomposed on our mind? For example, a waterwave's medium is water. Though there is an independent word to mean a wave vibrating air, which is sound, naming it as airwave or a wave in air might be no problem. Then what is the medium for sound to propagate? Right, air. It looks ridiculous but let's name light as lightwave. Then what is the medium of lightwave? Way to go, light. Which means we believe lightwave as light. What we experience is lightwaves, not light itself. And recall light is an electromagnetic wave. It means lightwave is composed of electric wave and magnetic wave. This will be discussed at the next chapter but matter is also a compound of electricity and magnetism. To put all these mentioned previously together, light and matter including our sensory organs and body are a product of the medium of light. It is like a figure in a painting is unable to figure out what he or she is painted of.

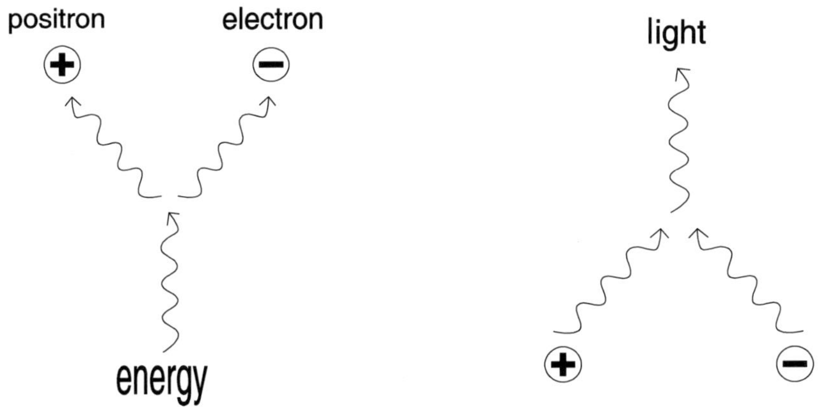

figure 1.5 a space believed empty actually lets out something.

The high voltage in the cathode tube induces cathode ray, which is a flow of electrons. They are from the negative terminal but not all of them. The high voltage drives electrons out of the terminal and the electrons scratch through the space which is vacuum. Meanwhile the scratch strikes pairs of particle out of the space like the depicts. And they crunch together to disappear right after they came to existence. If a vacuum has nothing in itself, what did the couple of an electron and a positron come from?

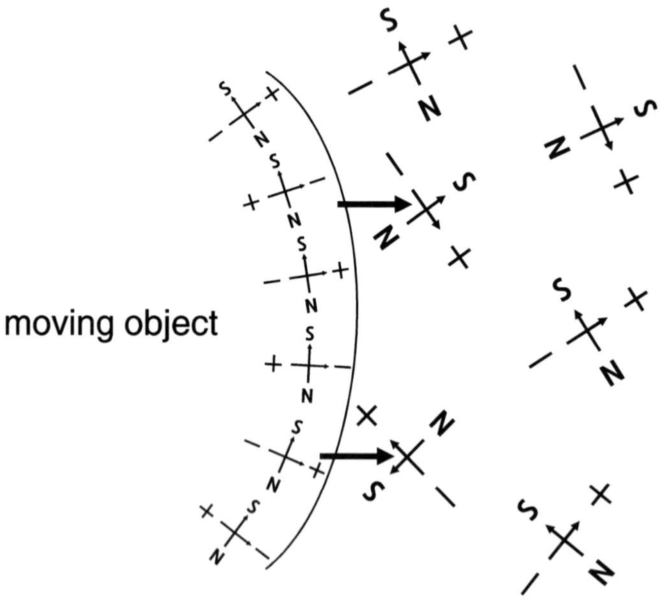

figure 1.6 heat is around an object, depicted as crosses of electricity and magnetism arrows, for heat is also a light, which is electromagnetic wave. The same is just in tangle in space, in structure in matter. This idea can be the ground of Brown motion.

The drawing above(figure 1.6) is a depict of a body in motion, which absorbs heat to gain weight. After acceleration of an object, its mass does not increase any more. The gain and the loss are the same amount when in the uniform velocity. Here is how it works. A body in whole is neutral electrically and magnetically. But they are not nanoscopically. Those polarized charges' moving is nothing but electric currents, and that so powerful like the ones of the cathode tube by high voltage. The motion of an object would mine pairs of particles out of the empty space just like what happens in cathode tube.

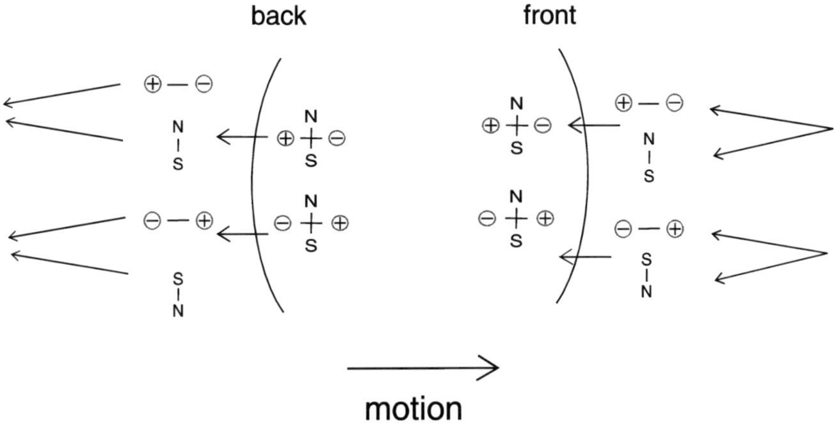

figure 1.7 heat literally breaks into electricity and magnetism by collisions and sticks together with charges in the moving object. As much as the acceleration, it accumulates, and flows backward out of the body

Those new lives in the tube were gone to be a flash of light in no time but they of a flying object stick into the body, because there are many opposite charges to hold them down in it. This is why an object's kinetic mass is greater than its rest mass. If so, what is that which jolts out of space? The answer is it is light or heat. Heat is said to be the kinetic energy of minute particles but it is also O.K. to call it just a light. Heat is so called infrared, which is also a sort of light. Heat is a tangle of light, while normal light propagates in a certain direction. This heat's properties is the cause of Brown motion.

The space is full of heat, which is a light. Light is a combination of electric and magnetic field as depicted. It gets broken into two elements when it collides with a polarized particle. After separation, they become a part of a moving body or return to naught with light emission when they have no object to bank on.

The phenomenon of light's decomposition into electricity and magnetism by collision is also the reason why there establishes the principle of uncertainty. For an object to gain a remarkable amount of mass, it should move so fast, which could hardly be observed. But the situation inside an atom is a lot different. An easy understanding is that the density of light inside an atom or a molecule is tremendously high, because all the charges insides radiate electromagnetic wave strongly. So an electron of an atom might look like the figure.

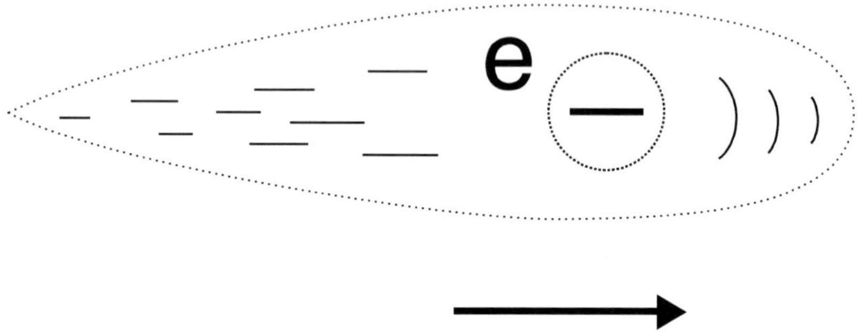

figure 1.8 electrons swell up and elongate as they get momentum in an atom where the density is extremely high.

The principle of uncertainty is that you can not determine its momentum and location at the same time. Look, its being itself has blurred. The momentum is mass times velocity. Electrons are in motion in an atom. As it moves, its mass and velocity vary so drastically.

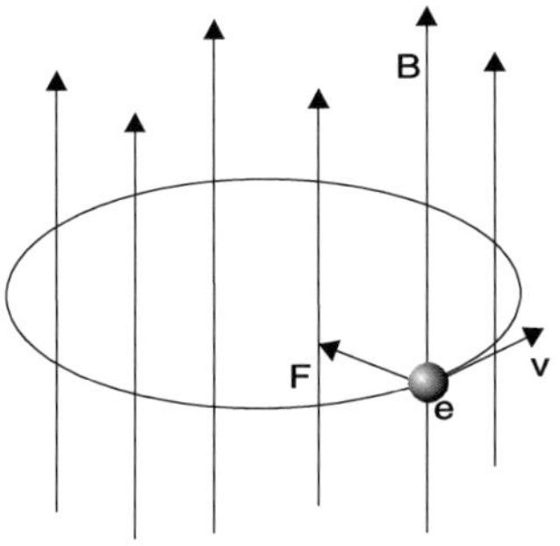

figure 1.9 Fleming's left hand law establishes for space full of heat.

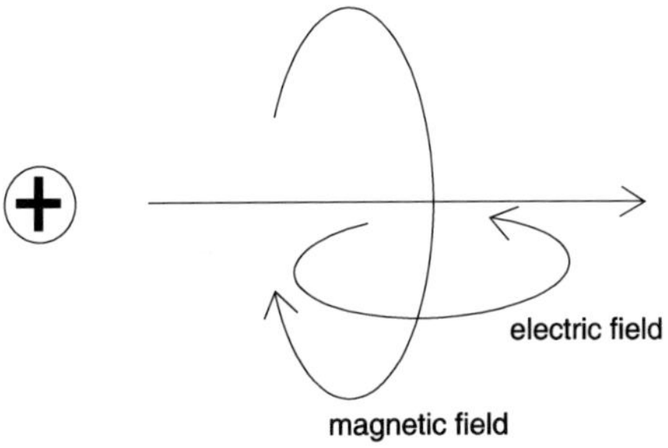

figure 1.10 electromagnetic wave transmission is also possible thanks to heat in space.

The principle of M and Yang covers beyond the boundary of matter. Nevertheless just taking into account light's existence filling the space brings us clues to many conundrums of physics. A critical point of Maxwell's equations is a change of electric field induces that of magnetic field as depicted above. But no physicist brings out a question why and how those electromagnetic phenomena occur. It is said James Maxwell had hypothesized a weird model of space, though. Furthermore, why an accelerated electron always curves to the left, not to the right, in a magnetic filed? Particles in a fluid vibrate very quickly and irregularly.

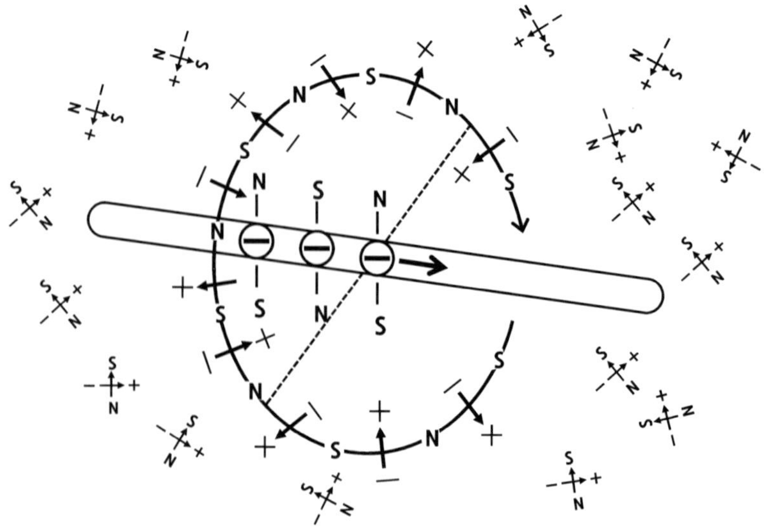

figure 1.11 Ampere' law is just the orderly alignment of strewn heat.

This is called Brown motion. And the cause of their shaking is heat. We just see the space as full of air but do not even imagine it is filled with heat, namely light. An electric current in a conductor is a linear flow of electrons. Let's take a look at how they flow one by one. An electron is regarded as a tiny bar of magnet as well. Electrons' influence is at rest when the current is at halt, because their electric and magnetic influences are canceled out to be neutral. Once they move, it induces the crosses of light or heat in the space to array in order. That's how such inductions take place.

A static electricity is stipulated as an influence by the early physicists. But why don't we remodel it a little more shrewdly? Influence but something springing out swirly. Furthermore we can postulate a thread of the magnetic or electric flux as a curling advance, not just a plain line. Then it gets us a picture of why there must be Fleming's left hand law.

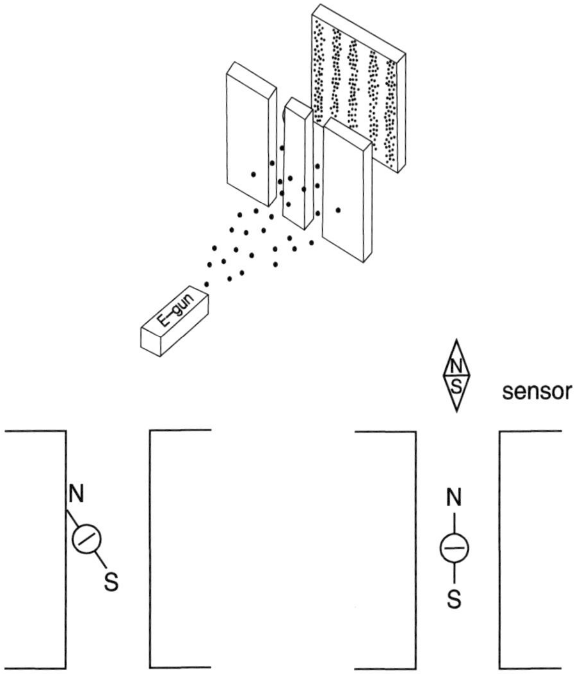

figure 1.12 an electron does not go through two slits at the same time. Just we do not know which one it has been through

That double slit experiment is an instance quoted frequently to explain what superposition is. Superposition is a state where a thing seems as if it exists at two places at the same time. It is said that light to have passed through the slits makes an interference pattern on the screen. Light can pass through both slits at the same time because it is a wave. But one thing regarded as spooky is random shooting electrons, which are a particle, also displays such a pattern. At this, physicists say an electron passes though the slits simultaneously or it has the properties of wave as well, as De Broglie insisted. He said all the matter are of particle as well as of wave. It is another pun as light is said so.

An interesting thing regarding the double slit experiment is the interference pattern does not take place when there are sensors installed at the slits to check through which an electron comes. So someone often comments on this phenomenon that the particle electron has an intelligence so that it can itself decide which way it will go, or when it observed, it chooses what state it will put on, of particle or of wave.

An crucial factor in the experiment is it has to be vacuum. With air in there, electrons do not draw the pattern. That it is in the state of vacuum means there is only one being that affect electrons' actions, which is heat. Marking heat as crosses like those in the picture is very convenient to understand what a flying electron looks like. An electron is an electric charge as well as magnetic. The direction of its magnetic axis is under influence of the polarities of heat around it. Thus an electron towards the slits would fly randomly changing the direction of its axis. Then those passing right in the middle of the slit would mark in the center of the screen and the others closer to the walls of the slit might veer more or less to the right or the left according to the angles formed by the axis of an electron and the wall surface. It was not a ghostly phenomenon. A shot electron has been definitely through one slit. We just do not know which one it is. An electron observed, it would not show up a pattern on the screen. To see which way it passes, you need to install a sensor which is nothing but a electric charge or a magnet. It will erect an electron's axis in a certain direction. The erecting prevents the angular momentum(rotating swirly) from taking place. Thereafter electrons do not draw any pattern but two files on the screen.

Now entanglement has also turned out to be not such an extraordinary event. One electron generated in couple, in Einstein's thought experiment, is taken from the other to the end of the universe. It says when one turns out up, the other down. What matters in this experiment is whether the spin up and down is decided at their birth or checking out what spin one of them is determines its spin and the other's. If the latter is true, how it happens? How they communicated with each other? Modern quantum mechanics takes the side of the latter and says it is proven by an experiment. One electron of the couple is parted several tens of kilometers from the other. They say it turned out that sensing one's spin decides the other's. So it is said the two are entangled. This event, however, is so ordinary to know the surrounding of the experiment. First of all, those electrons are separated through an optical fiber cable in which there is no air. It's the same environment as in the double slit experiment. There is also heat inside the fiber cable which is able to play a role of medium. When sensed, the electron has its axis erected. The sudden erection of an electric as well as magnetic charge causes a pulse in the space and the pulse is transmitted to the other with the information of up or down. And It would not happen if they were parted from each other in the distance of the universe. First, there must be no particle nor any strong wave in the way to each other. Second, there must be an adequate amount of medium in the way. Both Einstein and Bohr were ignorant of heat's cruciality even in a thought experiment.

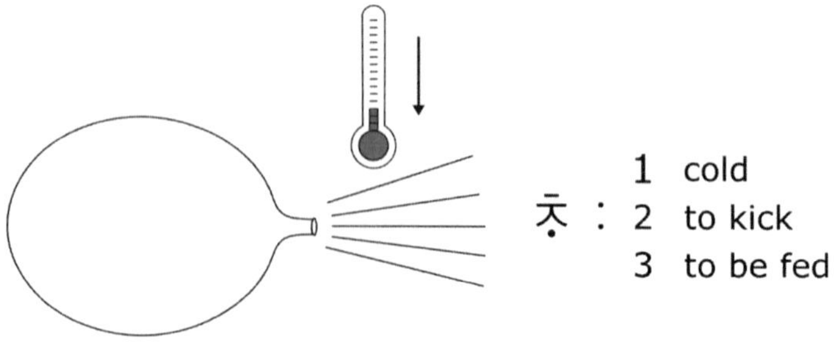

A spouting air is colder than other calm ones. Why? The expression of physics about air getting cold is the air absorbs thermal energy from around it. As a matter of fact, the air molecules just bump into heat divided and save it in the form of mass. So it turns cold. This is why a moving object emits heat when it collides with others to stop. A moving object has piled up heat from the front during acceleration. That Korean word CHa in the figure(1.13) has three distinctive meanings. But the three make sense from the depict.

2

The world woven by the sun, stars, and the moon

Most of stumbles of modern physics seem to have been generated from a wrong or unsuitable terming. What scientists can not help is they have to use those terminologies which were termed by the early scientists who were more ignorant of the study than they are. The most representative one is, I believe, the word 'charge'.

There has been the question what electricity or magnetism is since the beginning of electro-magnetics. According to early physicists, Faraday, Ampere, Gauss, and Maxwell' equations, electricity and magnetism are fundamentally not different from each other. But there is still no answer to the question what they are. Their first interest was about static electricity. Some material radiated invisible influence and they saw it as electricity. This is the beginning of a departure. No one had experienced a mass of pure electricity or pure magnetism till then. Of course, yet no one. So they happened to believe electricity lives only inside or on matter. Coulomb might probably have said the word during the experiment to deduce Coulomb's law. He charged a particle by touching it with an already electrically charged material. The early physicists' recognition of electricity, modern people and scientists' too, was just like Coulomb's charging a particle with electricity. There is a matter. There is electricity, with no idea where it is from. Charging the matter with electricity meant loading it with electricity. Like loading a vehicle with articles. But what scientists should wonder about should be 'Why is there not electricity only material?'

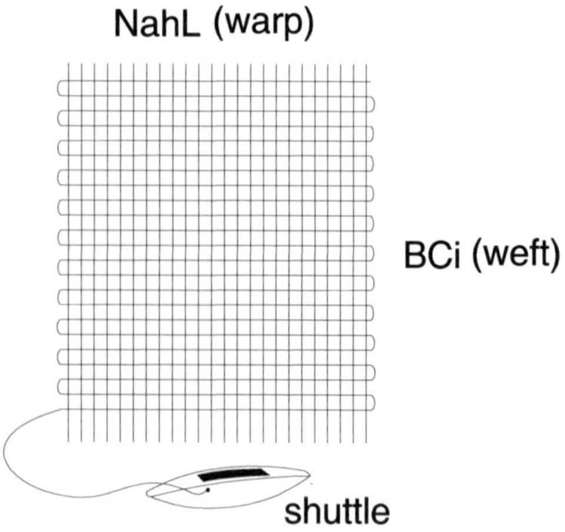

figure 2.1 a tiny bug would see the material as a jungle of crossed long threads. Particles, bodies as well as matter might be a recomposition of M and Yang.

Contemplating on weaving is pretty helpful to understand what matter and electricity are and their relationship. The writer's mother tongue Korean language is so intertwined with principles of M and Yang that a reference to it can be a sound guide to understanding how the world is framed. What we recognize of that woven material is a patch of cloth. Though they are made up of two bunches of warp thread and weft thread, when one was asked what it is, they would say, "That's cloth." The cloth or clothes is, however, not a real thing. It is a use reborn by human labor. The use actually exists on our mind. This is a good analogy to understand how matter is related to electricity and magnetism. There is cotton of which the weft and the warp thread are made, of which are woven shirts, skirt, pants, socks, underwear, and so on. Therefore the idea that matter is an independent isolated being and electricity may be charged on it is wrong, intuitively speaking. All the matters are just crafted up of electricity and magnetism. There is no other thing but M and Yang in the world. That is to say it's not that a particle is electrically charged, but that a particle has gained more electricity and magnetism.

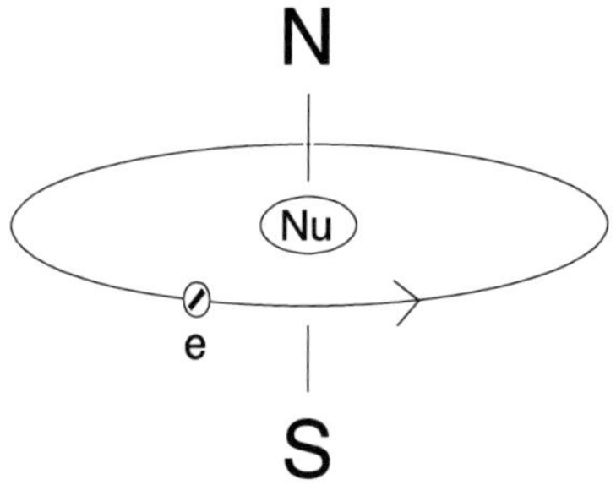

figure 2.2
modern physics' approach to magnetism is it might be something induced with electricity.

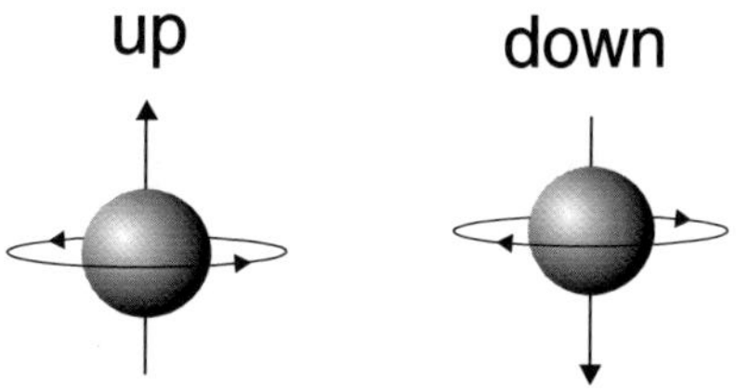

figure 2.3
an electron is too small to secure enough momentum to generate the amount of magnetism it registers

The scientists have been wondering where magnetism comes from, while the source of electricity was known, which is a tiny particle, an electron. Their flowing is an electric current. What became a clue for this curiosity was Maxwell's equations whose gist is changing electric field induces magnetic field and vice versa. They looked into a magnet to see what causes magnetism. Unfortunately there was no such thing found but an alignment of magnetic axises of atoms. The alignment was the source of magnetism. An atom had turned out a magnet too. And they looked more into an atom to see how an atom becomes a magnet. There was an electron orbiting around the nucleus. An electron's orbiting is itself an electric current, which can induce magnetic field. The next and last contemplation was on an electron. Interestingly, an electron is also magnetized. In other words, an electron is electrically a negative charge as well as a magnet. They wondered again where the magnetism came from. What they noticed at that was an electron's spinning. An electron has mass definitely and is a negative charge. Its spin seemed able to induce some magnetic field. But an electron was too small to generate sufficient momentum. According to their calculation, an electron should be bigger than its mother atom. Hopelessly it was the end of the journey to the source of magnetism. And it was how the electron's properties happened to be called spin. Though the magnetism is not from an electron's spin, it began being called spin just because it seems like that. The question what the source of magnetism is was left unsolved.

Electricity is not what magnetism induces. Magnetism is not what electricity induces either. Besides, both of them are not what can only exist charged in matter. They are two but all components to compose a matter just like there are only warp and weft thread to compose a cloth, even though there are numerous articles made out of them. There is a common maxim in Korean society saying, "Everything is made up of M and Yang." Though it is mostly addressed M and Yang, it is rather better to say M-Yang. Korean people also refer to it so. It implies there is nothing composed of only M or of only Yang in the world.

Just as the word 'charge' reflects misguided modern physics, so did another wrongly termed word, induction. The electric current through a metal wire seems like an only electricity. But as the figure, that is just a offset of a magnetic component. The same but of the electric component is conducted inside a magnet.

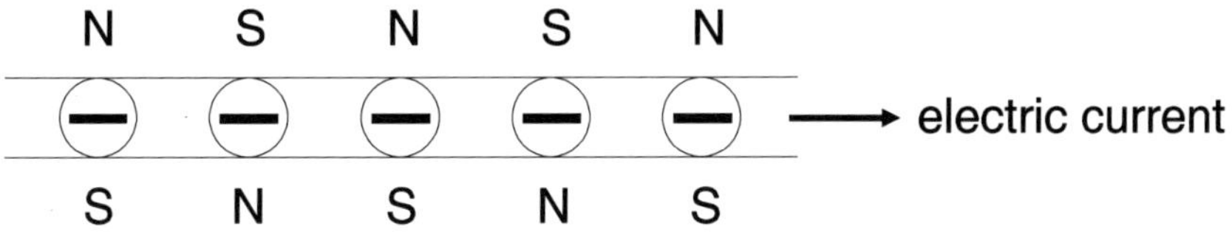

figure 2.4 it looks superficially like there is electricity only.
All are composed of the couple

As a user of clothes, we are aware of cloth, unaware of warp and weft thread. But to a weaver, those strings of thread are in the world of the weaver's perception. This is a good analogy to understand how the world is displayed. it seems like there must be previous states of only electricity and only magnetism behind the curtain. A weaver works on cotton to spin out strings of thread and dresses them into the warp and the weft. There are terms for only M and only Yang

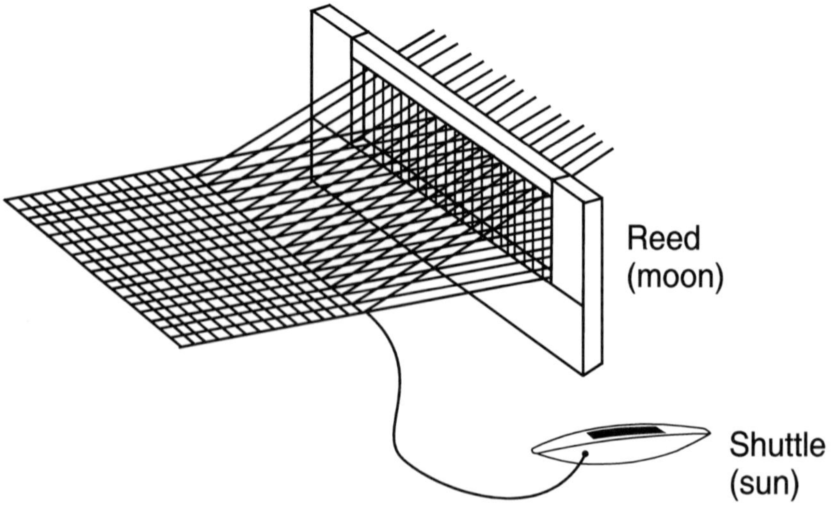

figure 2,5 all the beings are a product of NahL and BCi – warp and weft. Korean words, the sun, the moon, and stars are expressions of those celestial bodies' roles.

that are in pre-state. NahL in Korean language means the space up in the sky. And it is also warp thread as a homonym. It implies the space up in the sky is full of some material unknown to modern people just like the warp thread is laid out in the loom. And Yang of the pre-state is called BCi which means seed. Weft thread is BCi thread in Korean. No one knows what they feel like. Someone living in the world of sound is unable to sense the existence of air because he or she is also a sound. The world of NahL and BCi is not of knowledge but of enlightenment. Understanding it was the ultimate theme of religions. So was it at least ancient times.

A striking fact about NahL and BCi is that gravity is generated by NahL's action on matter. M and Yang says the reason an object falls is not because Earth pulls it but because the space NahL pushes downward. And let me open a far more stunning one, where the NahL comes from. That should be located above it because that causes gravity. Right, stars. Let's recall the weaving scene now. Numerous strings of NahL(warp) are already held down on the frame and weft thread is supplied between two columns of warp row by row. It is a refreshing picture that stars are NahL suppliers. Then who might be supposed to be the BCi(weft thread) supplier? Sure, it is the sun. The most terrible misunderstanding in understanding how the world we live in looks is believing the sun is the source of light. No, light does not come from the sun but is a compound of NahL from stars and BCi from the sun. This is why physicists could not identify the medium of light. NahL and BCi are not a material of this world before the third agent's action. So their beings are marked as void or empty in our sense. There are three sorts of celestial bodies in our linguistic world. The sun and stars and the last is the moon. So it feels like there must be a part the moon was supposed to play. The moon is a kind of a mediator or a catalyst. It puts together or connects NahL and BCi into light. By this moment there appear electricity and magnetism in the stage. There is a nice prop to show what the moon's role looks like. It is a reed, not a plant in the riverside but the device to guide the warp thread in the weaving. It beats a weft thread into the grab formed by a couple of warp threads. Thereby cloth weaves nice and neat. The moon is a reed in the sky.

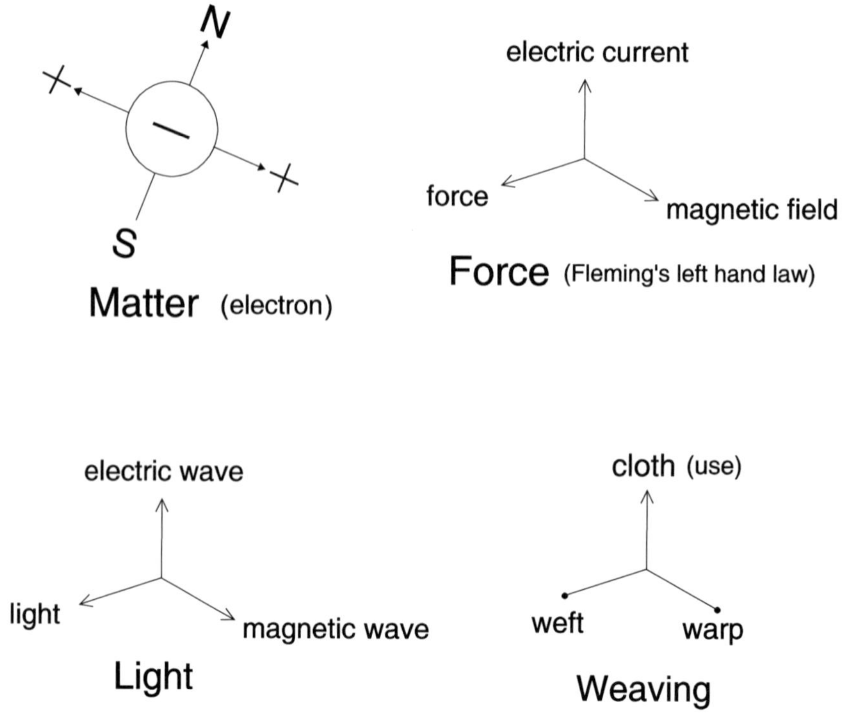

figure 2.6 This world we live in is designed. All the matter is composed of electricity and magnetism. They were NahL and BCi from the stars and the sun.

Look! Beings of physics are all in an identical formation. There is no such thing actually as a particle. What look like discrete things to us human are illusionary images. Even though an electron, which is regarded as matter, looks like it carries electricity and magnetism on itself, it is actually a product of electricity and magnetism. This means what an object's mass is turns out to be the sum of vectorial products of electric and magnetic field in a space. It can be likened to the area of cloth. We feel the mass as weight and the area as use. Definitely we are not the weaver.

3

Things to make Earth look round

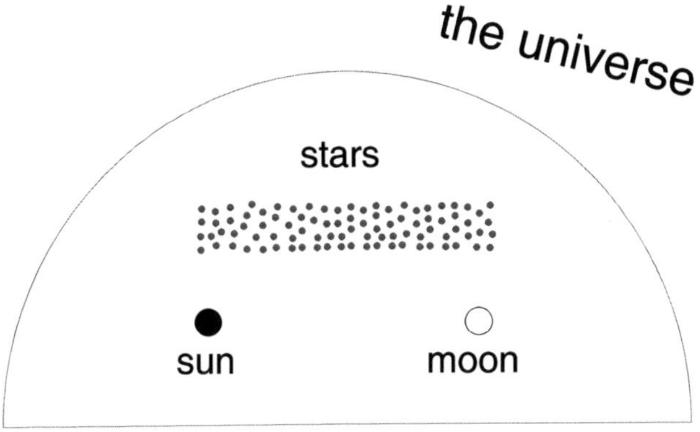

That light is a compound of two elements offered by the sun and stars signifies now it is time to have a new better picture of the universe. The East Asian countries, China, Korea, and Japan, call the universe Woo-Ju(宇宙). This word is composed of two Chinese letters which have an identical meaning, a house. They have had to make up a new faith that there must have been a profound philosophical reason of why their ancestors called the universe a house, since the western conquerors began to take up not only the land but also their spirit. Teachers and missionaries from the west enlightened them that the universe is not round nor a house but boundless. With the new model of the universe on their mind, thinking their ancestors believed the universe is a house was a profanity. But it is embarrassingly pitiful that just hundreds of years after having denied what had been accepted as a common sense for thousands years, they should make a resumption to the previous deserted one.

It is interesting both flat earth believers and round earth believers rely on the same means to prove they are right. It is light. However, there are some facts both of the sides are ignorant of concerning light. They manage their assertions in the strong belief that light advances straight. Their misunderstanding and ignorance begin at that. Light seems to travel straight, without an external impact. But the truth is light curves anyway. The reason light does not move straightly forward is also not by an external impact.

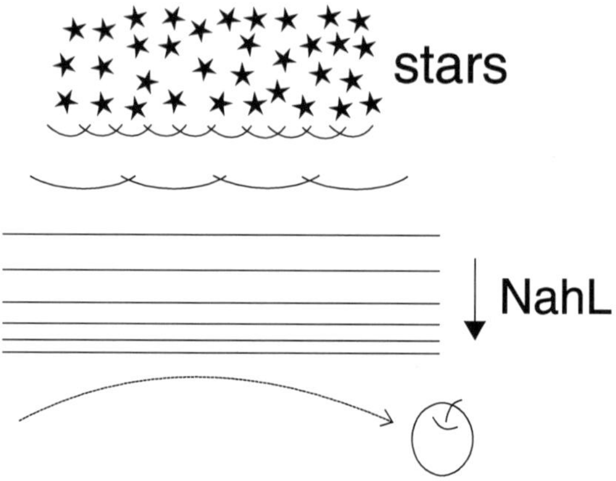

figure 3.1 NahL presses an flying object

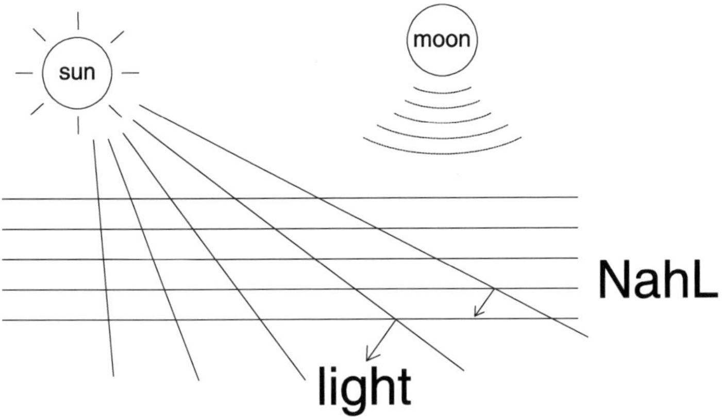

figure 3.2 light is not from the sun

First of all, stars radiate NahL(warp) and BCi(weft) from the sun is added to it. The mooning transforms the two into an electromagnetic wave, which is light. Its kinds, colors, are determined according to the angle formed with NahL and BCi when mooned. To express this effect more accurately, a determination of color of light is by the angular derivative. The one whose changing rate of angle by NahL and BCi is big becomes a short wavelength light. The rate is usually, however, in proportion to the angle itself. The bigger an angle is, the higher its changing rate is. So it is no problem to accept this concept as the drawings depict (fig 3.3).

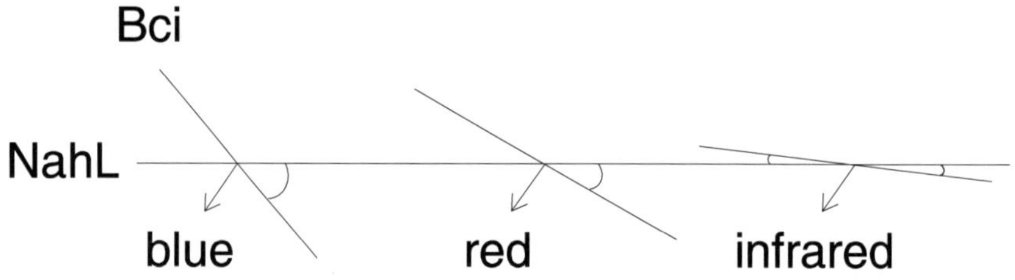

figure 3.3 Night can take place even in flat earth by this principle

By that reason light, which is ordinarily called so and looks like it comes from the sun, bends a bit downward when it propagates. The bending is not so remarkable in a short distance but so is that of a light to start from many tens of kilometers away. On the other hands, lights from stars draw a different sort of trajectory. The logic of the reason is so simple. NahL is descending, and that so fast.

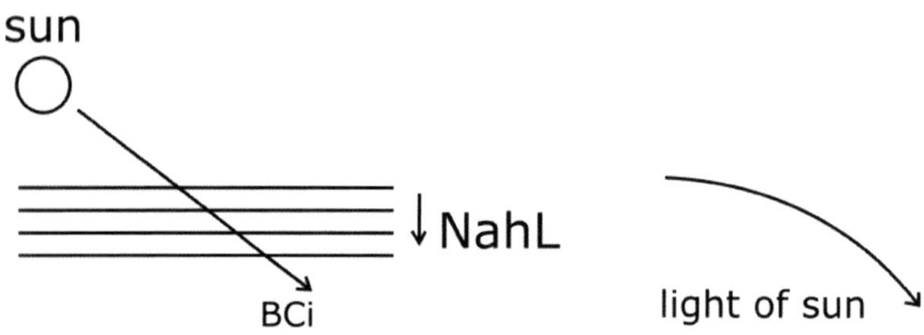

figure 3.4 light curves like a thrown object

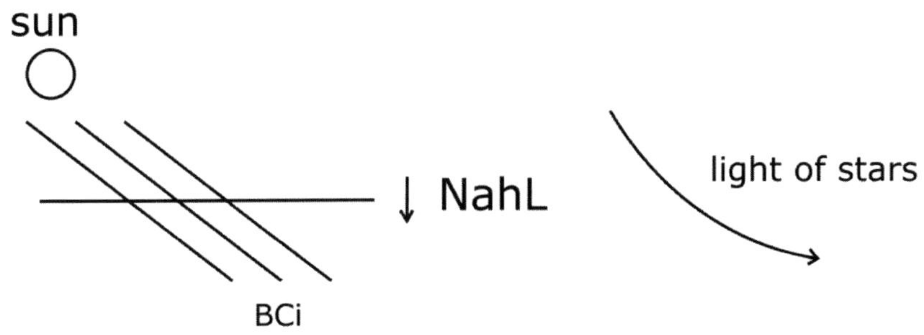

figure 3.5 light of stars curve like a 1/x function

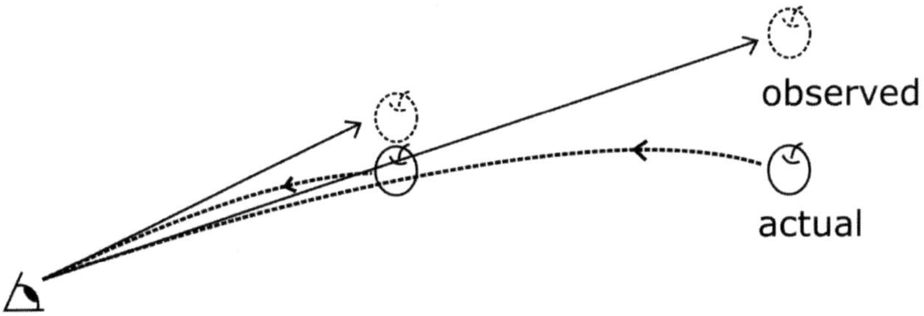

figure 3.6 An object quite away looks located higher

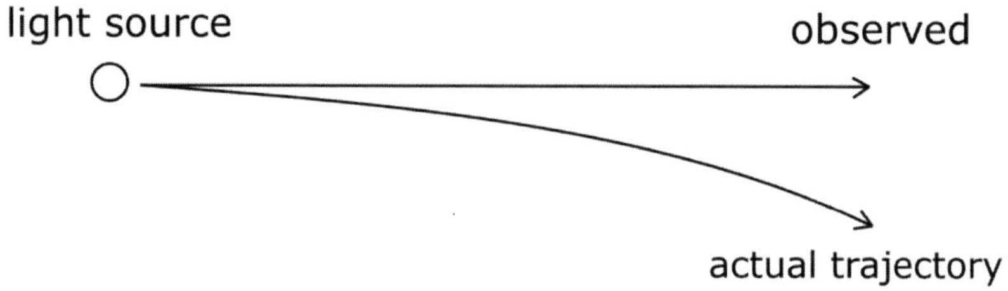

figure 3.7 a light looking straight curves down in fact

And there must be a question why it looks straight to observers and in a taken picture. This can also be explainable. Paradoxically, light curves so that it looks straight. The light from a farther light source looks like it comes to the observer from more above than the source really is.

To apply this phenomenon to the scene that a ship approaching the land shows up its peak first, now we come to the understanding. The ship looks behind the hill of water or below water. This is why when zoomed in, it shows out its hidden bottom part. In other words, it's because light curves that Earth looks round as it is actually flat.

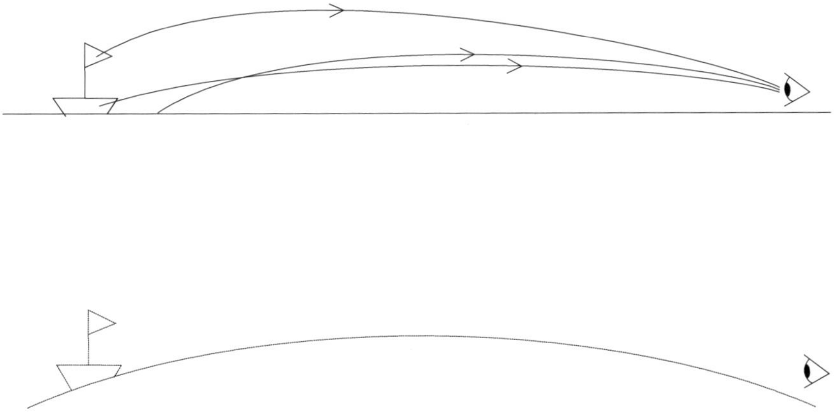

figure 3.8 a ship looks like it is beyond the hill of ocean water

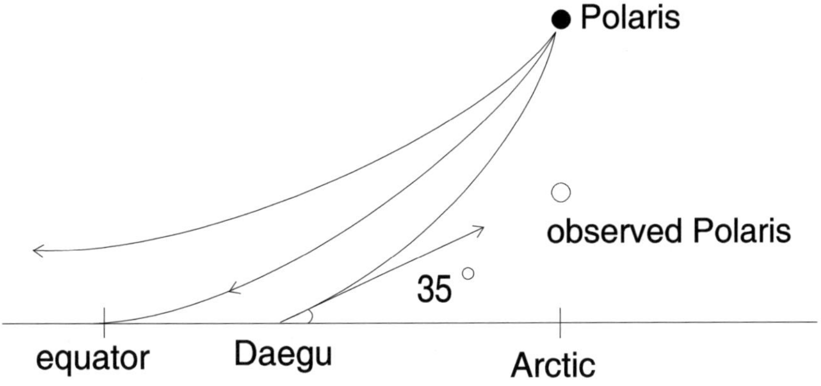

figure 3.9 altitude of Polaris at my home town

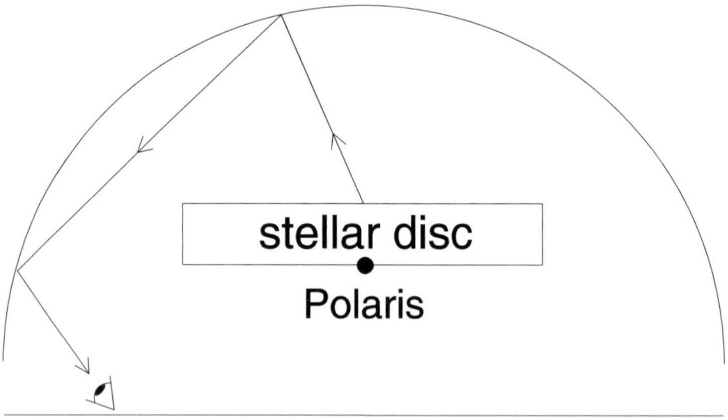

figure 3.10 Stars in the southern hemisphere are from the upper side

Realizing lights from celestial bodies curve their own ways leads to the answer to why even in flat Earth Polaris' altitude varies along places and people in the southern hemisphere are unable to see it. The star light just flies over the area of the southern hemisphere because it curves like the figure 3.9.

The stars you see in the southern hemisphere are reflections of those in the other side of the stellar disk(star disk). Though it does not seem to make sense at a glance, that system is so crucial for the sun to reciprocate between the Tropic of Cancer and the Tropic of Capricorn. So the fact that the direction of the rotation of the stars in the southern hemisphere is counter to that in the northern hemisphere can not be an evidence to prove Earth is round.

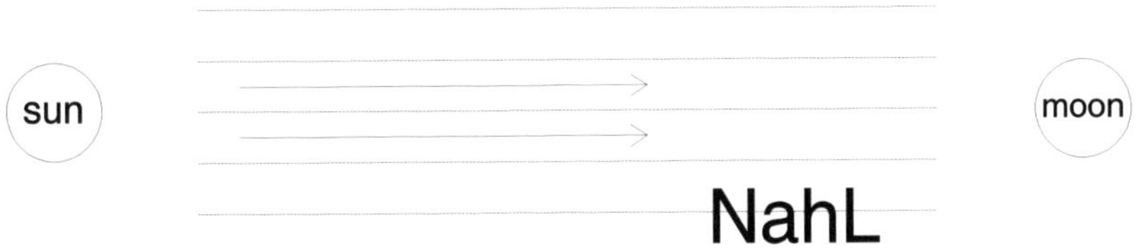

figure 3.11 dark when BCi and NahL are in parallel or in a small angle.

A lunar eclipse is Earth's shading on the moon. But what if the shade is not Earth's? The picture taken of a lunar eclipse shows two different features from that of a solar eclipse. One is that the round shape is not a clear cut as of a solar eclipse. In other words, it blurs. The question is why it has to? The other is that the moon is shaded black but soon red and back black, which is not observed in a solar eclipse. As depicted in the drawing(fig 3.11), the moon can have a dark round mark on itself. The acceptance that light is generated from compounding of M and Yang, in other words NahL and BCi, is a precondition for this idea to make sense. To postulate a color of light or a kind of light is determined by the angle NahL and BCi form, a light of which the angle is less than that of infrared would be black to our eyes, that is to say, a dark mark when it lands in a surface like the one in the moon of an eclipse. And the reason it turns red and back to black is explainable. Red is a sensible light closest to the critical point between visible and invisible. The value of the angle formed with M and Yang varies back and forth at the critical point, as the sun and the moon orbit along their own path.

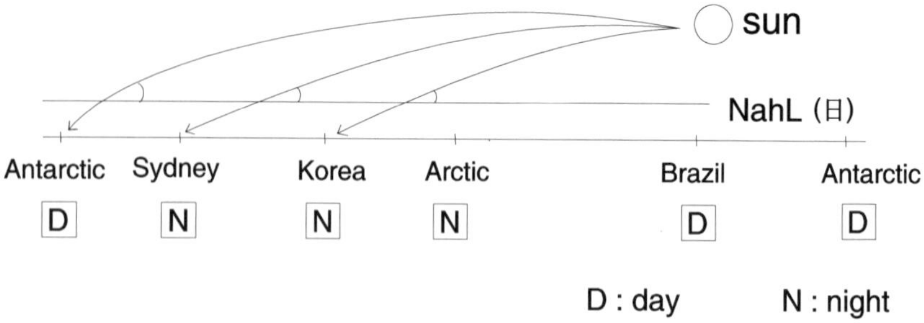

figure 3.12 a profile of white night at Antarctic

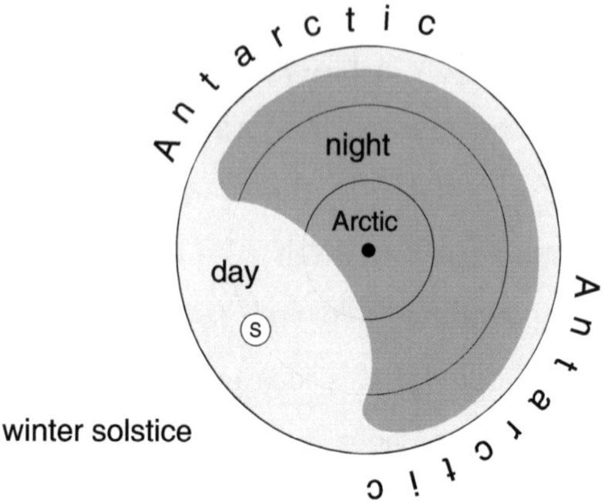

figure 3.13 view from above

One of those which seem impossible to take place if Earth is flat and possible in round Earth is a white night at the Antarctic. But this one can also be overcome with the fact light curves and its colors are determined in the angle of M and Yang. The reason a night takes place even though Earth is flat is the angle M and Yang forms when the sun is gone far away from the observer is so small that visible light flies over the area where the observer stands and only invisible light, for example infrared, shines. But a place way much farther from the sun than the ordinary places, for example the Antarctic area when it is winter solstice, can rather be bright for twenty four hours a winter day(of December and January).

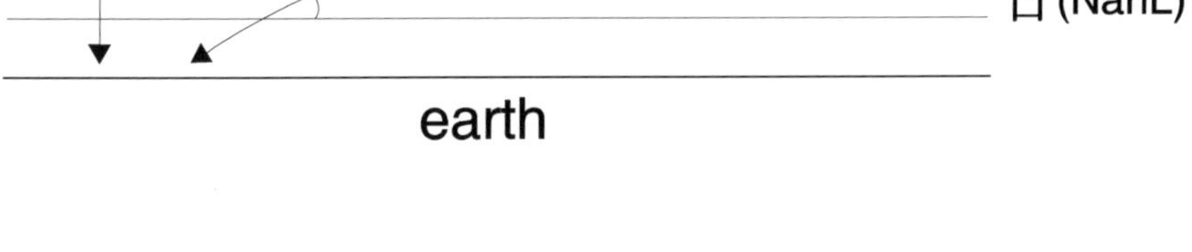

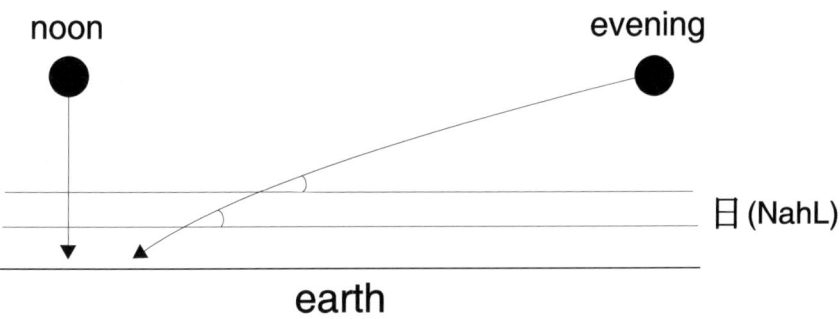

figure 3.14 it's a sunset, not earthturn

The way a ship looks like it fades away beyond the hill of ocean is nothing but how the sun looks like it sets down beyond the land. It's not because Earth is round but light curves. However there seems to be another problem standing in our way to believe Earth is flat. The sun always look almost the same in size.

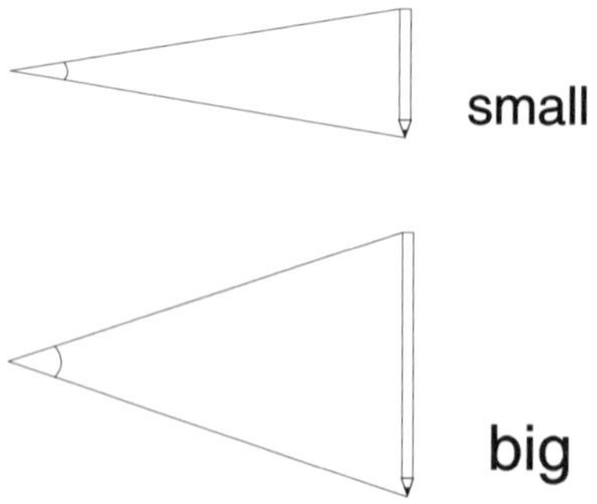

figure 3.15

In a flat Earth, the sun in the morning should be located much farther than it is at noon. It's got to look different in size along the time. This problem can also be solved with the consideration of what it means looking big or small.

Looking big means the angle formed by the couple of an object's terminals is big. As depicted, the light to leave from the top end of the sun travels a longer way than that from the bottom end. Thereby the angle that forms in an observer's eye shows up rather bigger than that which straightly advancing lights from an object would make.

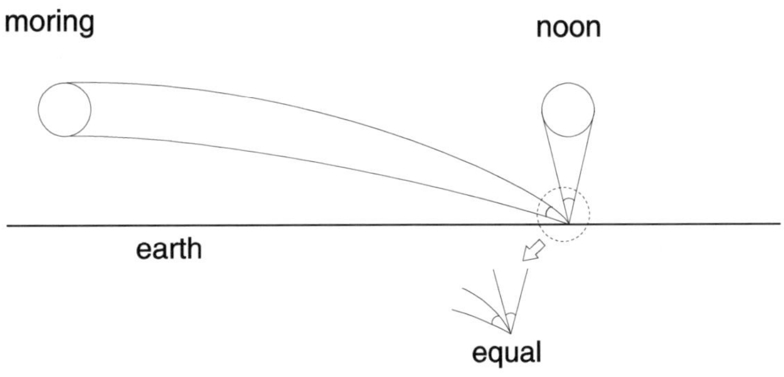

figure 3.16 a light from a upper part curves more.

In spite of the correction about why the sun looks the same in size, there is still a further way to go. That has explained only why the sun looks vertically the same size all the time. It has got to look so horizontally.

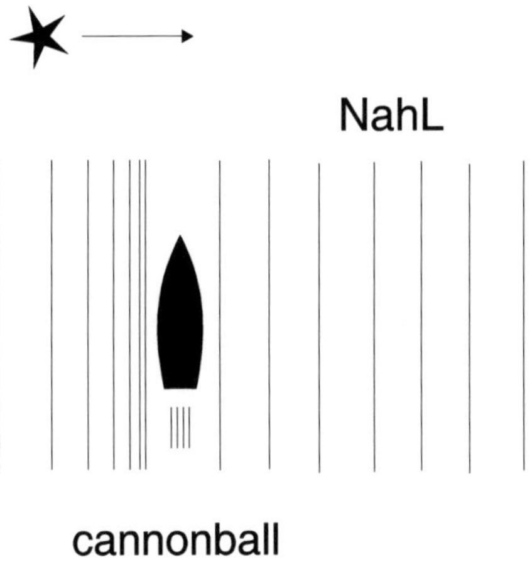

figure 3.17 Coriolis effect east west component. NahL gets denser on the east side as stars rotate.

So it is necessary to put the horizontal component in shaping the sun. Ones tend to mention Coriolis effect or force as an evidence of the rotation of Earth. To explain this effect briefly, a flying object has a tendency to land at a point slightly shifted from where it was supposed to unless Earth rotated. But the effect does prove on the contrary that Earth does not rotate. If it does, the effect must not occur because the turning effect of Earth' rotation is added to the flying object when it departs the surface of Earth. Then, what on earth generates the effect? The answer is no other than stars. We have learned stars produce NahL which is the origination of gravity. If stars rotate, they will cause a thin heap of NahL on the east side of a flying object. So it gets a little budged. This is what Coriolis effect really is. But there is still an embarrassing thing of the effect. The explanation makes a good sense only for an object flying from the north to the south or vice versa.

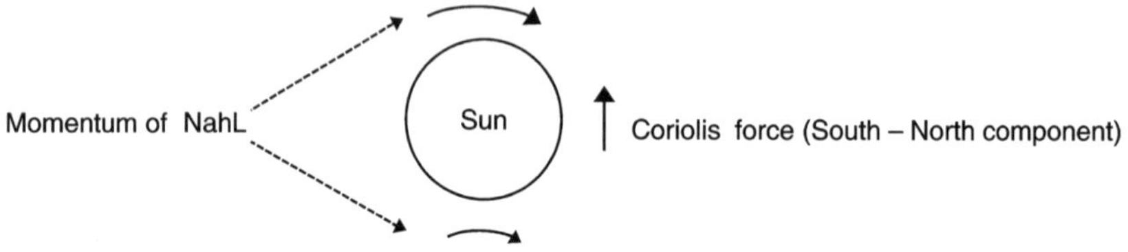

figure 3.18 the sun reciprocates seasonally under Coriolis effect south north component

What about it flying in the direction of east and west? Embarrassingly it shows an identical effect to that of an object flying east to west.

Hydrodynamics says that velocity difference of two streams causes a force. The lift of aerodynamics by which an airplane is able to lift upward is a good example. The air above the wing flows faster than below. It causes the space above the wing low pressured and below high pressured. High pressure means density of air at the point is high. The density of NahL is almost uniform all over the sky. But it sets up a different aspect when an object is up in the space. As stars above rotate east to west, NahL also gains a horizontal component of momentum. The stars in a more outer side rotate a longer path per hour. The NahL on an object's south side has more momentum than on its north.

This is no more than the feature of the lift of an airplane. The force shows up, whose direction is from the north to the south. It was mentioned that light curves downward in the sky. While Coriolis effect being explained, it has happened to issue a derivative knowledge. There appears a density difference between the south side of the sun and the north. Therefore light curves horizontally too and the angle of its width like that of its height becomes bigger than that of light in the supposition that light flies straight forward. This is all the story of how the sun looks the same in size all the time.

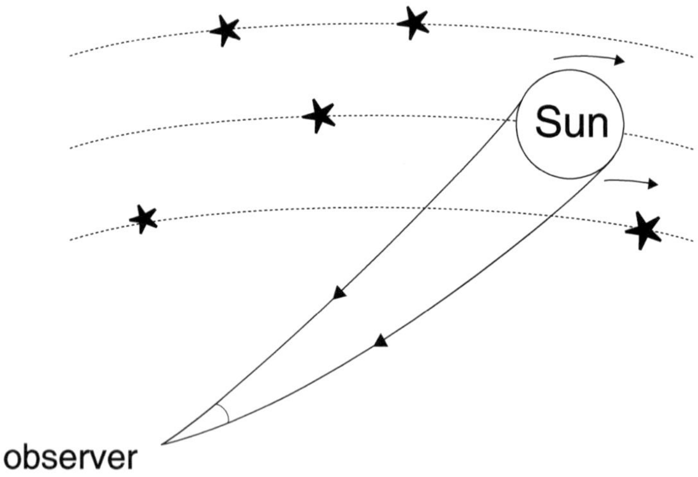

figure 3.19 the sun looks identical in width too

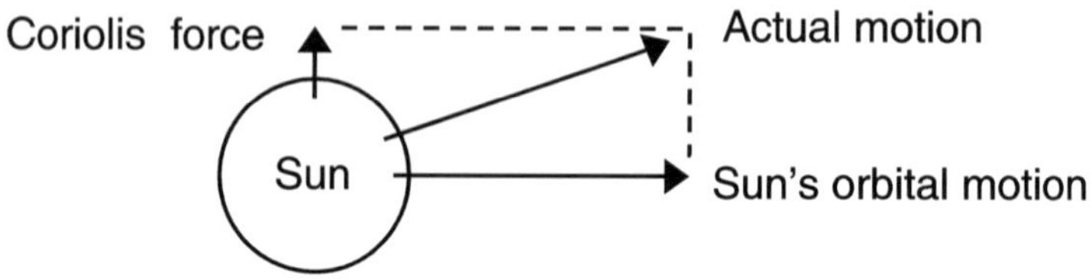

figure 3.20 the sun gets faster toward winter solstice

The grasp of how Coriolis effect takes place is so beneficial. The sun does reciprocation between the Tropic of Capricorn and the

Tropic of Cancer. For the sun to perform such a motion, it needs momentum. The sun is a flying object in the sky. So it can be postulated that it is also under the Coriolis effect. Since the sun orbits east to west, it seems free from the east-west component of Coriolis effect. But it can be affected when it gains the momentum toward the south, which is generated by the density difference between both sides. As a result, the sun orbits steadily faster, slipping toward the south day by day. That is to say, the sun spirals down to the south for winter and up for summer. And the density stand switches the other way since the sun crosses the equator. This picture of the sun's seasonal operation is possible thanks to the premise that the NahL distribution of the southern hemisphere is by reflection of the upper side of Stellar Disc.

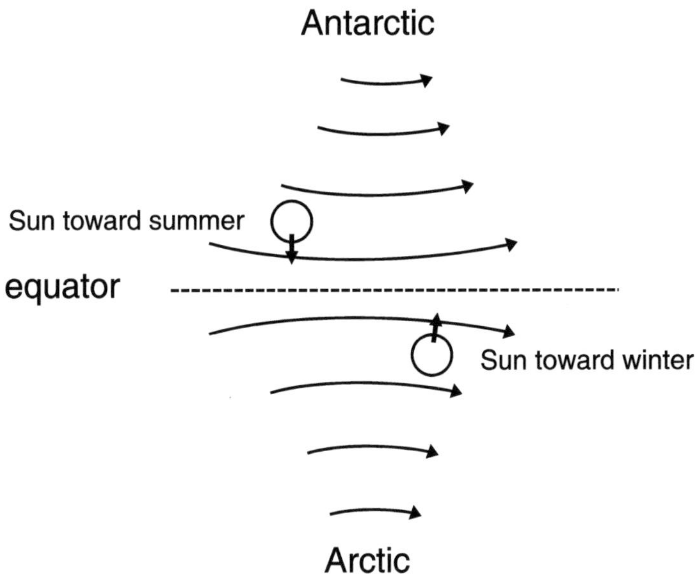

figure 3.21 the sun's reciprocation

4

Heaven is round and Earth is square (天圓地方)

The title of this chapter is how my ancestors think the world is. No, what they taught their descendants to believe. So had done not only they but almost every one before the western science began to misguide people. One I'd like to add about this maxim is that line is not only about the geometrical shape of the world we live in but the operating mechanism of the world. It is rather mathematic.

The beginning of electro-magnetics was not smooth. The early scientists such as Faraday, Gauss, Ampere, and James Maxwell, had to face and got to solve the question what on earth electricity or magnetism is. It is so important to define terms and concepts when a study begins. We modern people are so used to its being-around that we believe we know what it is. Electricity is everywhere nowadays. But can we answer the question to what electricity is? Neither did the fathers of electro-magnetics. The first hypothesis was that electricity is a fluid which comes out from the positive charge and disappears through the negative. It sounded good but also had very serious problems. A grain of dust can also be a electric charge, negative or positive. From where it appropriates all the fluid which spews for ever, without any influence that can change its electrical polarity? In terminology, it violates the law of mass and energy conservation. This issue was quite a controversy at that time as well and called out serious disputes. At the end, they had to be content with a compromised version that electricity is not fluid but influence to be depicted as arrowed curved or straight lines. And the nature of electricity, identical polarities repulsing each other and distinct attracting, was also regarded sort of as a definition. As definition means why they attract or repulse each other was not found and is not yet, but will. The scientists at the time just defined that electricity is so in nature. But the line, "Heaven is round and earth square" recited in the Eastern society since thousands of years ago, can be an answer to such natures of electricity as well as magnetism.

$$\Phi_E = \oint_{\text{surface}} \vec{E} \cdot d\vec{A} = \frac{q}{\varepsilon_0}$$

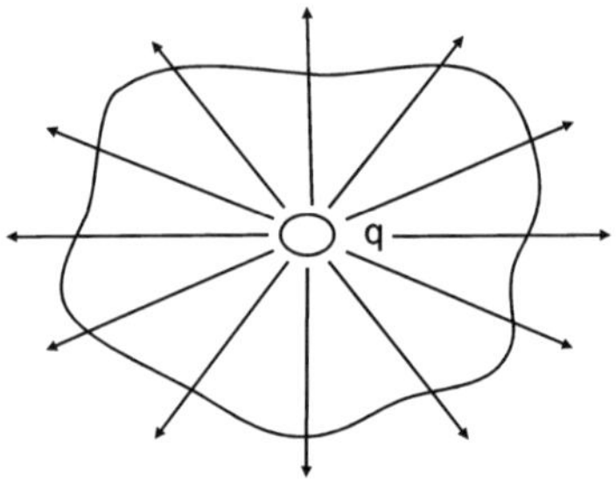

figure 4.1 Maxwell's first equation

figure 4.2 where that electricity comes from?

The physicists of 19th century concluded electricity is just influence, not fluid. But in reality, they handled electricity as if it is fluid, not to mention modern scientists. An example for this trend is the first equation of Maxwell's. What the equation stands for is the total electric flux radiating from a static electric charge is equal to its charge. what is flux? They say electricity is not fluid but they calculate the amount of electricity as if it is water from a spring. What's wrong with them? So I, who is a descendant of M and Yang founders', employed my imagination to knock down this problem.

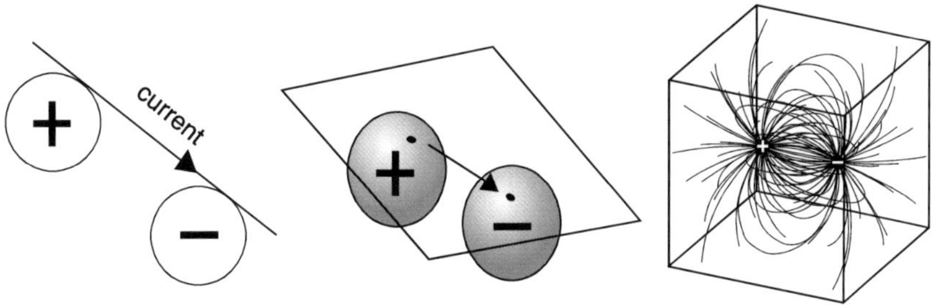

figure 4.3 electricity from the other world?

I thought to myself when I read the early scientists' logical predicament that electricity might be something like what is depicted in the figure 4.3. It is also OK that it is drawn on a single sphere or a circle. I'd like my readers to track down the geometric logic of the depicts. As the round figures start with a circle, advance through a sphere and up to an invisible four dimensional sphere, it grows from a tangent line, through a tangent plain, to a tangent cube. Just connect a battery to those round bodies, and on the tangents we can see what electricity looks like in this world. We live in a 3D world. So the third figure is the reality where we are.

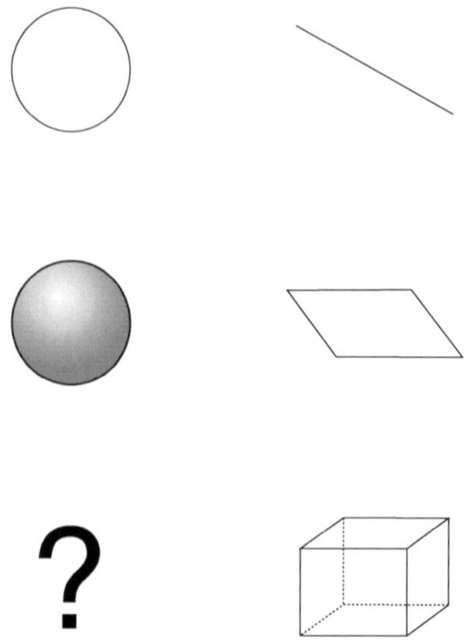

figure 4.4 there must be a heaven, though invisible.

The point is it's no problem even though electricity is a fluid, only if there is a world behind the curtain, which can keep supplying the fluid into this side where we are. It feels to us human like nothing exists around an electrically charged particle or object in the air. Time of believe or not again. But look! Why don't we get in a creature's shoes who lives in a two dimensional world like a plain slate, namely the tangent plain(the second figure of fig 4.3). It would wonder where that electricity came from, ignorant of the sphere on the other side, just as the early physicists who lived in a 3D world did. That invisible more than 3 dimensional world or sphere was called a heaven by ancient wise men, who crafted Korean language. And the tangent cube is an earth. You should have been asking, "Is there only electricity in the world? How such a figure of the world can be possible?". The answer is "Yes, there is only electricity in the world." Of course, there is only magnetism too. But why don't you read Maxwell's equations to see electricity and magnetism are fundamentally not different? Nothing else but electricity and magnetism has joined to form this world. And electricity is Yang and magnetism is M. But electricity and magnetism are things on the earth. They were NahL from stars and BCi from the sun before they transform into light, which is a electromagnetic wave. It is light when the angle of electricity to magnetism is 90 degree and matter when less than 90 degree.

And this saying is also a common sense in the East Asian society, "Every thing in the world, physicals and metaphysicals, is made up of M and Yang."

The motive person for me to do this work to reveal scientific secrets was Niels Bohr from Denmark. He was a fan of Taich, Tai-GG by his majesty, namely M and Yang. Even though he opened a new era of physics, he seemed to have had quite a dissatisfaction with his theories. He bumped into a Tai-GG pattern one day and happened to believe that the concept of M and Yang, which is two opposite ones living together, might be able to harmonize many contradictions of his quantum mechanics. He called the idea complementarity. But I can dare to rule that his idea about M and Yang is licking the surface of a watermelon or a blind man feeling an elephant on the leg. I am sort of a native of the society full of M and Yang culture. Anyway, let me fix the atomic model he suggested and won the Nobel prize for. It is said that he went to China to learn M and Yang. I am quite sure it did not pay so off. If it did, my this work should be unnecessary.

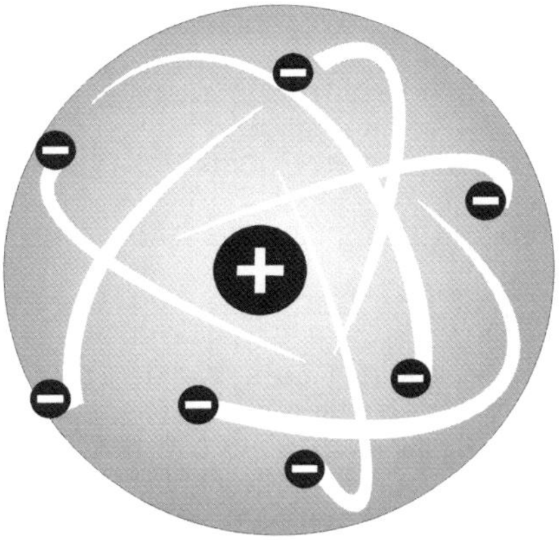

Lutherford's modle

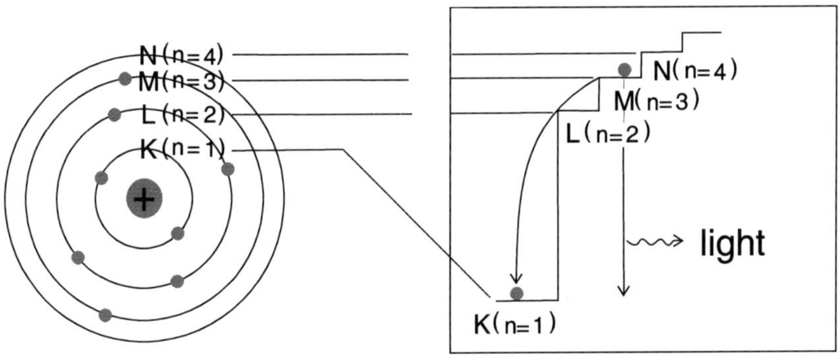

Bohr's model

figure 4.5 in fact, two models have the same problem unovercome.

His feat was that he suggested a better model to cover the shortcoming of Rutherford'. Rutherford discovered an atom is not a hard ball. There is a nucleus in the center and electrons fly quite far around it. The problem of his model was it fails to explain why electrons are not pulled down to the nucleus, as an electron is a negative charge and an nucleus positive. Niels Bohr's idea was electrons do not orbit arbitrarily around but in an certain level just like the depict. But it is just finding a fact, not the answer to the question why electrons do not fall to the nucleus. Finding this answer is very important. Because it is the key to the question what electricity is. In physics, a negative charge attracts a positive and repulses the other negative. But it does not explain how. It seems like the process to look for such a thing is itself physics, though. Let's check out how this takes place first of all.

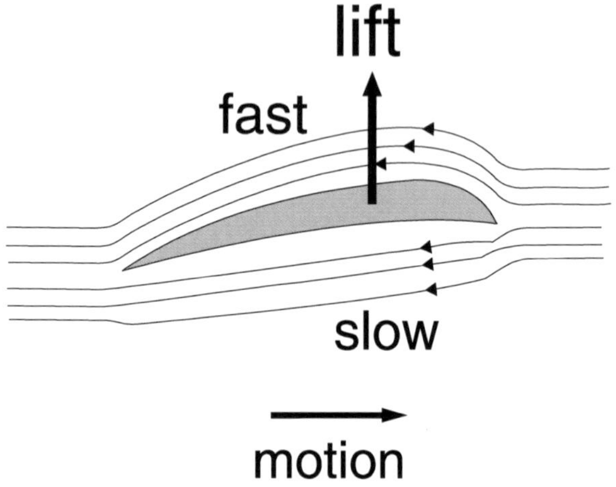

figure 4.6 lift of aerodynamics

The depict is how an airplane gains a lift. In virtue of the feature of the wings' shape, the speeds of air above and below the wing happen to be different. It flows faster above the wing than the below. The place where fluid moves faster turns to be a lower pressure than the others. And force takes place from a higher pressure to a lower. Thinking of electricity and magnetism as fluid helps unriddle many inexplicable secrets, for example, how a negative attracts a positive.

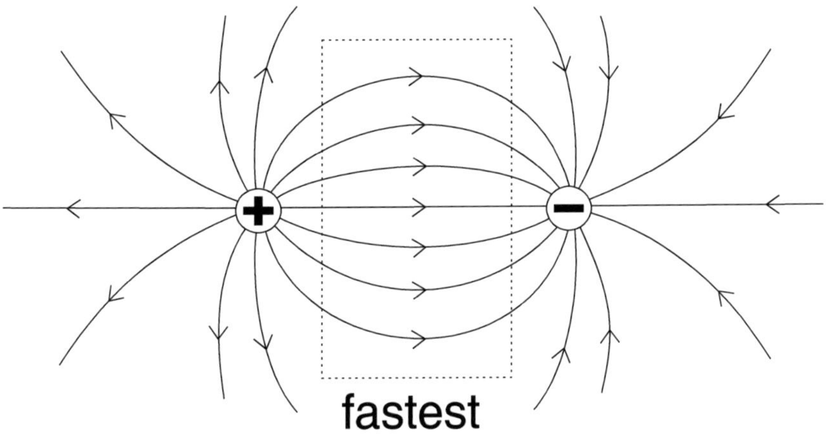

figure 4.7 electricity fluid idea explains how a positive and a negative attract each other

It was mentioned that the direction of a force is from a higher pressure to a lower. Have a look at where the pressure is lowest. Right, the space between the two charges, because electricity fluid flows fastest there. Then what happens? The two particles are pushed towards each other. This is the simple mechanism of how different polarity charges attract each other.

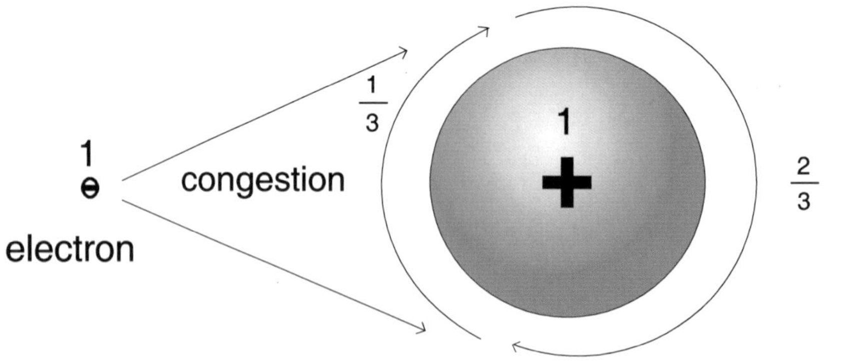

figure 4.8 big size difference prevents opposite polarities from attracting each other all the way

Have you ever thought why electrons are so small, while a nucleus or protons are huge, compared with an electron? Let's recall the scene an electron and a positron mass up together and disappear right after they are generated with high energy impact in the cathode ray tube. The question is why they can not last long. The answer is their sizes are equal. Likewise, an atom is able to keep in shape, thanks to the asymmetry in size of electrons and protons, or a nucleus.

If their sizes were identical, they would suck up each other to naught. But in that asymmetry quite a congestion takes place between an electron and a nucleus because the nucleus can not suck all the electricity fluid up when the electron has got close to the nucleus. Make sure it was postulated that electricity is fluid or something. Only a fraction of the surface of the nucleus is open toward the electron. Thereafter the fluid speed there would be getting lower, which causes a relatively higher pressure between two objects, small and big. The high pressure would make the electron pushed back away from the nucleus, but not all the way away. The electron would begin to be pulled back in when the nucleus secured a sufficient space to digest the congestion caused by the electron's approach. At last it turns out that an electron orbits around a nucleus, fluctuating. Modern physics has said an electron does not orbit in circle nor in oval but appears and disappears discontinuously. That is, however, just an observation. An electron shapes up large and long like a comet when it flies around in an atom. What we can possibly imagine is they(electrons) just look intermittently hidden or screened behind one another's tails.

The atomic model just remarked in this paper is a very good example of the spontaneous symmetry breaking. And this is very easy to understand. The sizes of an electron and a proton, which are electrically symmetric to each other, are so different. It's symmetry breaking. Breaking means changing of a state. But it seems atoms are ever so. The point is here, that matter is a product of electricity and magnetism. What we see are derivatives of functions with respect to time. The amounts of mobilized NahL and BCi to form an electron and a proton are asymmetric. Therefore the term 'spontaneous symmetric breaking' needs correcting. It should be 'intentional asymmetric designing'.

This hypothesis, which is a faith to me, points out that Coulomb's law needs fixing or abolition. The force is not always in reverse proportion to the square of the distance of two charges. Why this kind of malfunction happens? Because that law had been scripted before physicists obtained the knowledge of what's going on inside an atom. It is like a line of sentence from the declaration of American independence. All men are created equal. It seems the writers of the declaration were unable to foresee a time to come when male and female are believed equal in political rights. In the same context, the universal law of gravity can not help suspicion either. Both of them were born from a particle oriented view of the world.

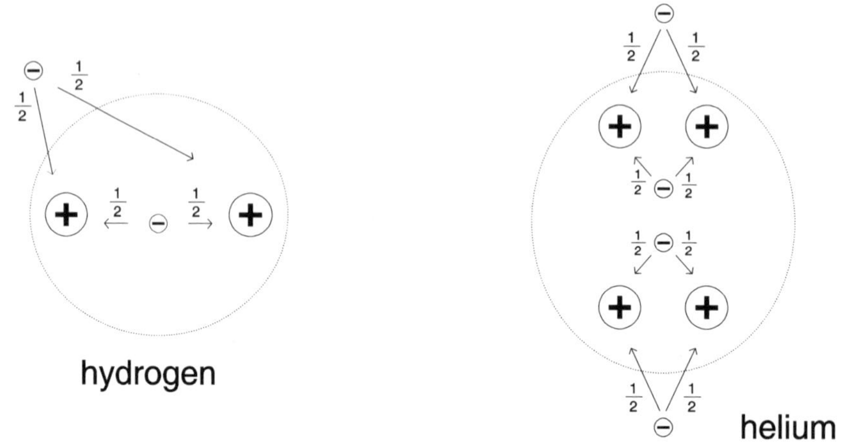

figure 4.9 with electricity fluid idea, gluon turns out redundant.

With the acceptance that electricity and magnetism are fluid, strong nuclear force is not real as modern physics argues but it takes place in the same principle in which an electron is prevented from being pulled all the way to the nucleus. There is one proton and one neutron in a hydrogen atom. One neutron can be separated into one electron and one proton. This electron can play a role in holding the two protons in to form a nucleus. The two protons' inhaling potential is, however, still alive. But an electron in the outer orbit would offset it up to a mechanical balance. And take a look at the figure of a helium atom. There are two protons in the nucleus. It was regarded as a wonder in the physics society that identical polarities can abide in that packing space. They were supposed to push each other so hard, according to the common sense of physics. It's such a basic. Thereby a particle was contrived to make this phenomenon sense, gluon. But such a thing is unnecessary in this suggested model.

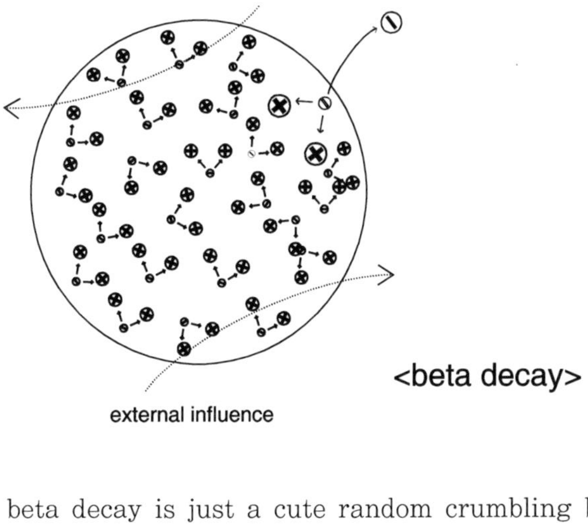

figure 4.10 beta decay is just a cute random crumbling by impacts from outsides

Beta breakdown occurs unpredictably. A new concept and a particle, which are strong nuclear force and gluon, were made up to cover up the wrongly oriented physics law, which is a plus charge always attracts a minus charge. So was weak nuclear force. The key point of this problem is why Beta decay takes place randomly. Beta decay is a neutron turning into a proton after an emission of an electron. This phenomenon is usually observed in a very heavy atom, in which so many protons and neutrons live together. So the structure should be very vulnerable. An electron's breaking off a neutron is, therefore, not by an unknown force but just by accidentally being jerked off along mechanical environment changes in and out of the atom. In the end, they all are just electromagnetic force. Here is a more accurate term, performance of M and Yang.

5

Heaven, Earth, and Man

A big handicap of modern physics is it employs two distinctive principles to account for one world. One is the theory of relativity, which is applied to talk about big objects. The other is quantum mechanics, which is devised to understand what is going on among nanoscopic beings. That two principles are used in one world is not problematic itself. But it feels like there goes something wrong, though this sort of inference is rather intuitive and somewhat superstitious. The beginning of misguide by not only those theories but modern science is a scientists' as well as other human beings' particle or object oriented viewpoint. Points, particles, and objects are nothing but a virtual, expedient, instrumental use by the designer of the world. They look real on our minds.

A stunning fact, of course this is an area of faith, is that looking real on my mind is not a natural happening but someone's cust -omization.

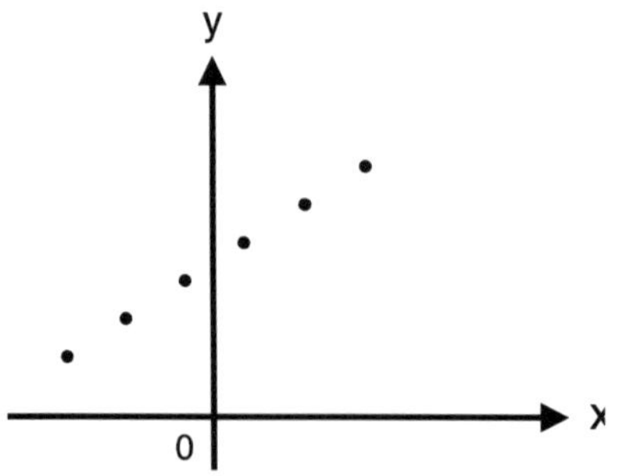

figure 5.1 the key point is the function. In other words, relations. Markings(particles in physics) are expediences.

There is a point put up by two straight lines, which is termed the origin. And there are many other points. Mathematicians use that coordinate to inform readers as well as themselves of the location of a point. Believing as a scientist or a philosopher that there are actually electrons, protons, neutrons inside an atom is helplessly naive. Religiously or philosophically speaking, such people of childish insight into the world are called animals, beasts, or a multitude. What is Descartes' intent of the coordinate? - it is said Descartes invented that system. Students should try to understand the relationship between the marked points. But fools study the axises and shapes of arrow tips and sometimes what the paper and the ink are made of.

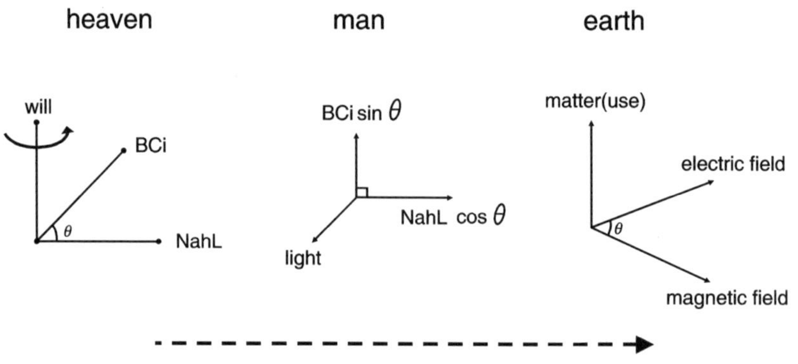

figure 5.2 a depict of information process

Analogically speaking, this world is a painting painted with two colors of watercolor, M and Yang. If really so, there must be a painter or brushes as well. Someone who is sufficiently spiritual would appreciate the work. Others who were not would be obsessed with the watercolor. It has been taught in the cultural society of M and Yang that three persons involve in the painting, heaven, earth, and man. This philosophy is so well reflected in the vowel system of Korean alphabet. Thus it can be inferable that consonants of Korean language are devised to express the features of M and Yang.

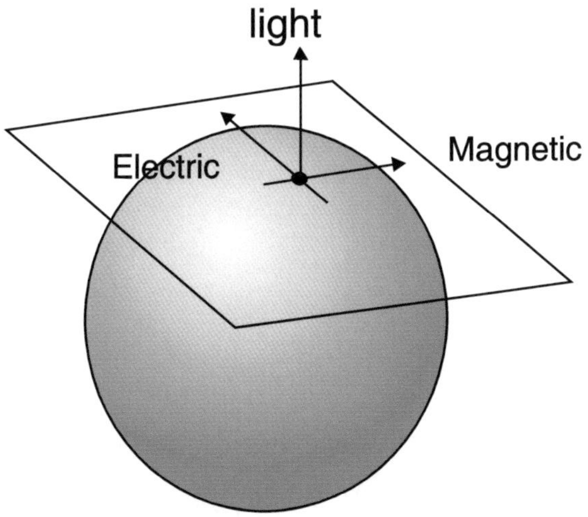

figure 5.3 the world we are experiencing is a series of derivatives of functions. To mathematically interpret the heaven is round, it is a circle or a sphere of at least 4D. It can be described with sin and cos functions, which can be differentiated infinite times without being zero.

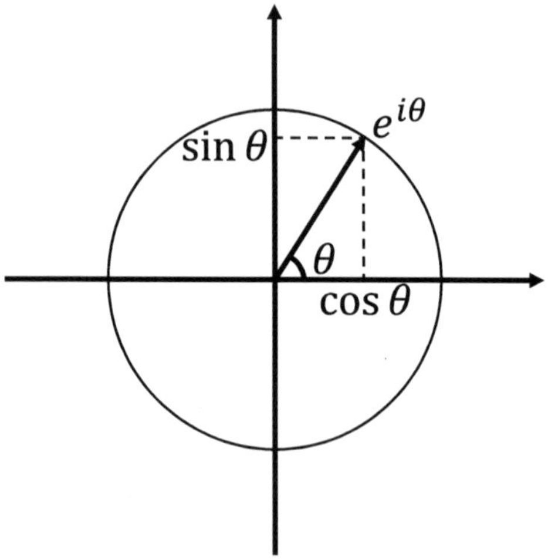

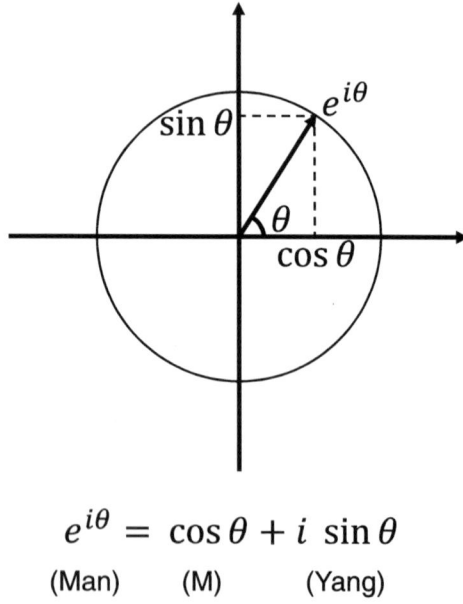

$$e^{i\theta} = \cos\theta + i\ \sin\theta$$

(Man) (M) (Yang)

figure 5.4 Euler formula is a good manifestation of philosophy of M and Yang as well as of three agents.
To quote the ancient scripture, "God is one, but the uses are three."

That picture is how light generates. NahL and BCi is a material in the state of superposition. We do not know what they are, nor even if there they are, so that they do not exist to us humans. The points in the coordinate have been there ever since. They become existences when marked in ink. The marking is what the moon does, rather like the reed in the weaving. This signifies a lot. There must be three agents for light to come to existence. Make sure where particularly the moon is in the universe designed by modern science.

How they can measure the width of the universe? It is said to be 15 billion light years. What do they mean by light years without light? No moon, no light.

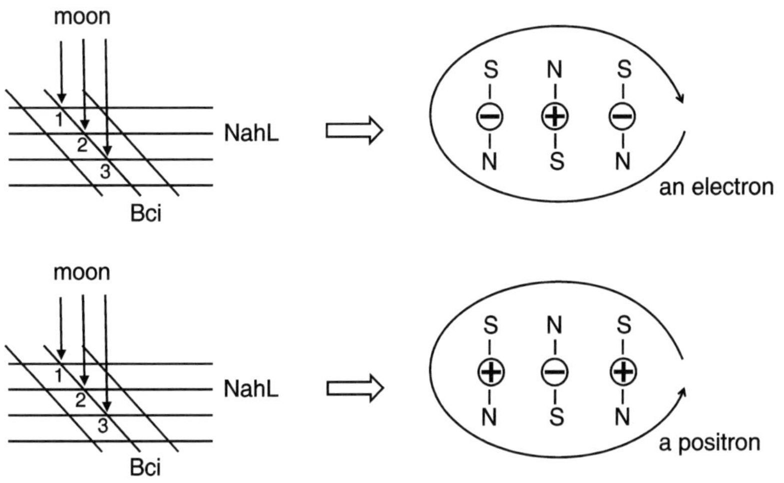

figure 5.5 by now it might make sense that particles, objects, things, whatever, are something like computer graphics.

The purpose of the coordinate is to understand a function. A connection of several crosses of NahL and BCi one way around is an electron and a conglomeration of far more crosses the other way is a nucleus. The electron springs out a bunch of mooned M-Yang and the nucleus sucks in or offset it. To avoid collapsing together to disappear, they must look a lot different in size. Thereby they are able to keep a distance through forming some congestion between them, as mentioned in the previous chapter. The congestion is the atom's mass.

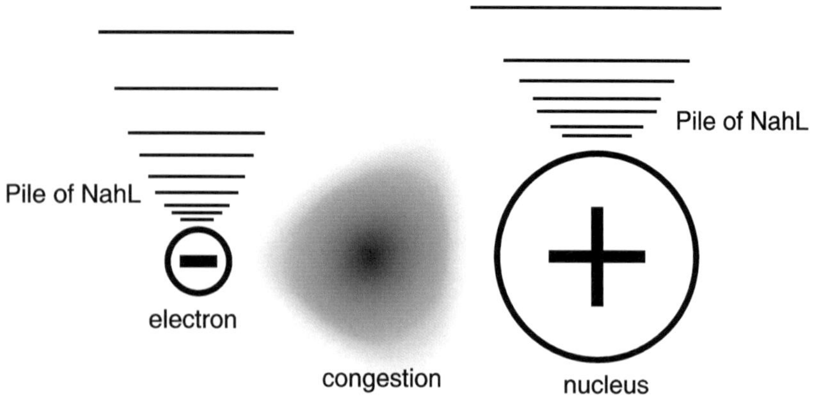

figure 5.6 why physics has to have two principles, quantum mechanics and relativity theory? all the forces are performances of M and yang.

The congestion of electromagnetic material(light) inside an atom causes a slowdown of NahL and BCi's materialization by mooning. Just like Coriolis effect, relatively high density of NahL generates a certain amount of force. This force upon matter(a particle) is the very gravity that had the apple fall. It is known that mass of an atom is that of electrons and those of protons and neutrons.

Then how that congestion can be the mass of the whole particle! The point is the particle model is just an expedience for our understanding. More honestly, that suggested structure of an atom is not real. It is like a coordinate system. That look of matter is also a display run by the universe operator, no matter whether you may see it or him as a computer or a ghost. Einstein is said to have commented after a severe discussion with Niels Bohr, "The moon really exists only when I watch it?" But the answer is it does never exist whether he observes it or not. What he see is what is shown by, well, it or him, who senses all beings' intentions and sends information to the observers' mind. As an individual, an atom or a proton also has a congestion inside themselves so that they can have mass in such a structure as above. And inside a proton is there the same structure as well. It's an endless loop. This is why the theory of string is meaningless either. The strings must have an amount of volume, which makes me wonder what is in them. Just make sure, whatever it is, what we sense or understand, even what we imagine is a show. We must have met the same structure even inside a quark, if we had tried to see inside it. The endless loop is also a shown to our consciousness. There is no particle but perceptions as such. So the congestion is a reflection of the mass of particles in the model under the postulation that those electrons and nucleus are points without volume.

The fact as important as how gravity generates is that matter is also of NahL and BCi by the sun and stars. Such a shocking announcement. Celestial bodies produce matter. And it was a common sense four thousands or five thousand years ago. They knew things that recent smart scientists are ignorant of. The writer of this paper was accidentally born in a society of which the language has kept ancient secrets.

It seems that almost all the debates over whether Earth is round or flat end up with the artificial satellite issue. This issue is crucially important to both sides. The existence of an artificial satellite seems impossible in the flat Earth model. So it is also understandable the zealots of flat earth believers deny their beings in the sky. Without their actual beings up in the sky, it turns out easy and convenient to draft a flat Earth model. I also wished so because this issue is not what I am able to physically check out. I've got so terrible an airsickness. But there is still a hope. There are the sun and the moon floating around up in the sky. Satellites might be able to fly the way those celestial bodies perform the flight.

Most of all, for a spacecraft to stay in such a high level of orbit where there is no air, it needs to overcome gravity or elude the effect. Either way does not sound feasible. But what if there is no gravity way up in the sky? Gravity is not an absolute value. It was a mere reaction to the congestion inside an atom. If the congestion is the cause of gravity, we can get rid of it by easing the tension. Matter in a extremely low temperature tends to turn numb to gravity. Some flows upward against gravity and another runs through the wall of the vessel that contains the matter. The secret how satellites can float high lies in those phenomena. Matter near the surface of Earth lives surrounded by a heated space, which means any space whose absolute temperature is not zero, namely about -273.15 Celsius, is heated. The heat around an object and the congestion inside an particle are essentially identical material. Being surrounded with heat means the object is under a kind of pressure from the outside. What if the pressure went down? The tension of congestion inside an particle or an atom would be alleviated. That the pressure of heat is low is a low temperature, in other words. This is why an object in a very low temperature grows gravity free. If it is cold enough in the high sky, there must be no gravity. Even though it is not absolute temperature zero up there, without air which is able to hold heat more effectively up to an object, the gravity free effect will be significantly strong.

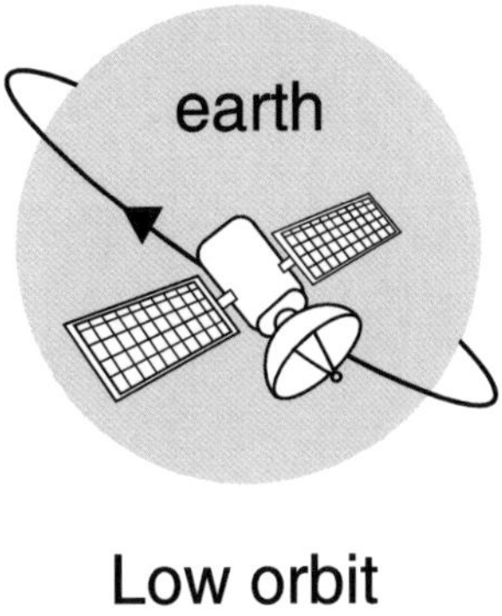

figure 5.7 a satellite orbiting in the postulation Earth is round

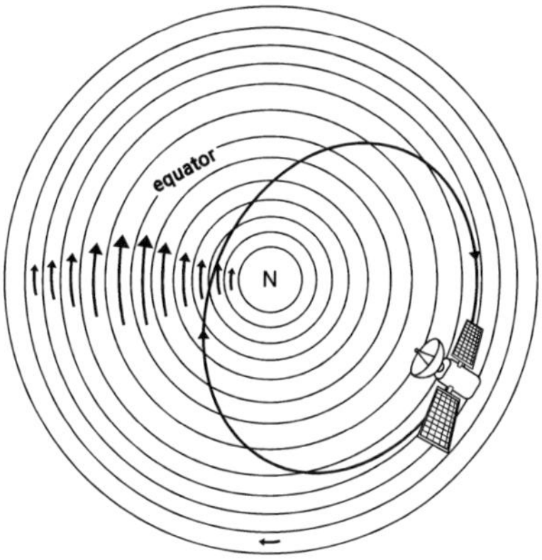

figure 5.8 in the postulation Earth is flat

There are two types of satellites that act against common senses. The stationary satellite seems impossible to orbit in the flat Earth model. Upon round Earth it actually orbits. It just looks stationary. It orbits as fast as Earth rotates. But it is not easy in the flat Earth model for a spacecraft to hover like a hawk in the sky where there is no air. To make it worse, stars rotate and cause Coriolis force on the satellite, which slides it to the west every moment. What matters is how to keep it from shifting to the west. But there seems to be no Coriolis' effect in the equator, which is like a transition belt. It would be no problem if a satellite was let to live there.

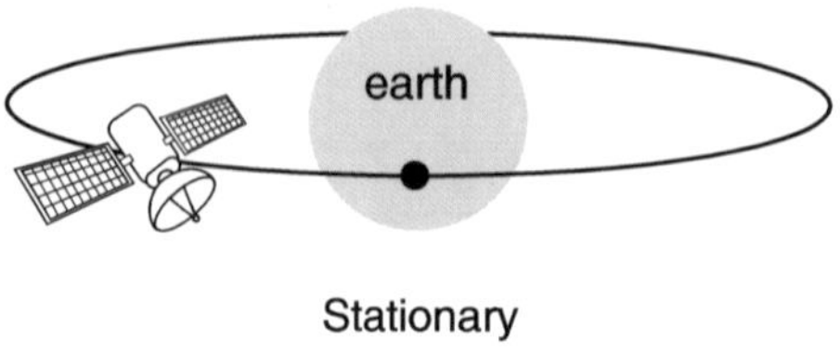

figure 5.9 this satellite moves, in fact.

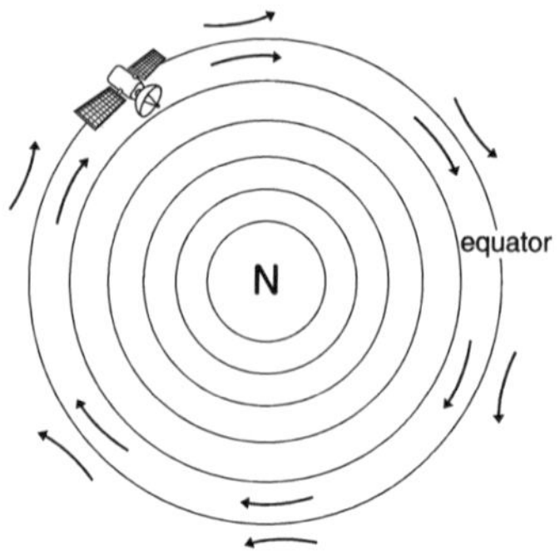

figure 5.10 Upon the flat earth should the satellite not move.

Another riddle to solve is on a polar orbiting satellite. This seems to stink flat Earth believers the most. In round Earth its motion is a single way rotation. But the satellite in the flat Earth model reciprocates between two sides of Antartic. It would be, however, no problem with the newly announced knowledge through this paper, of how matter is born. The roof or dorm of the universe is often postulated to be a hard glasslike material. It could be not. Imagine an air bubble under water. There is no cover nor a membrane to separate air inside the bubble from water outside. There might be no such mantle between inside the universe and outside. Therefore another imagination is possible. What would happen if a polar orbiting satellite collided into the boundary of the universe?

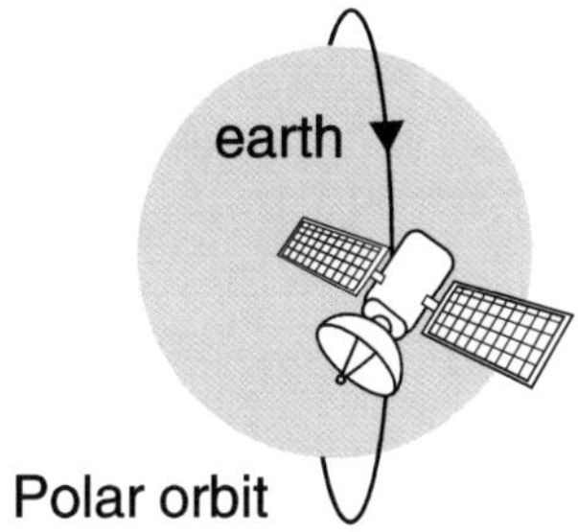

figure 5.11 satellites in circular motion on round Earth

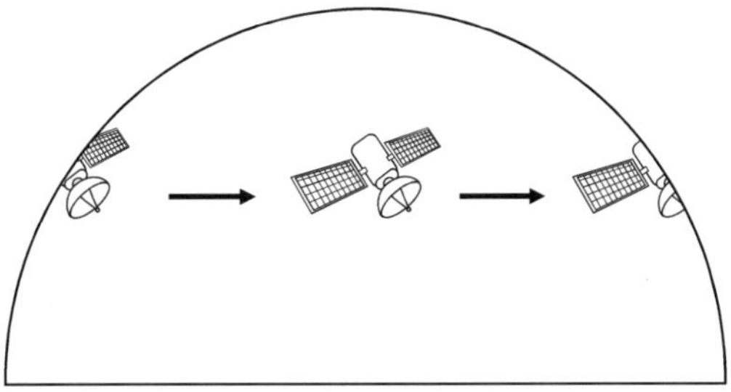

figure 5.12 satellites in reciprocation in flat Earth, but.

I know that's a story of Harry Potter. But think what matter is. It is our percept drawn with the sun, the moon, and stars. Matter does not exist where they do not perform. And the thought or belief that an object disappears this way at the Antarctic and reappears the other way takes to the idea that the universe or its operator may have intelligence, perhaps emotion too. A character in a computer game disappears when regarded to have gone into a virtually existing house. We all know it's just an information manipulation, not a real disappearance nor a reappearance.

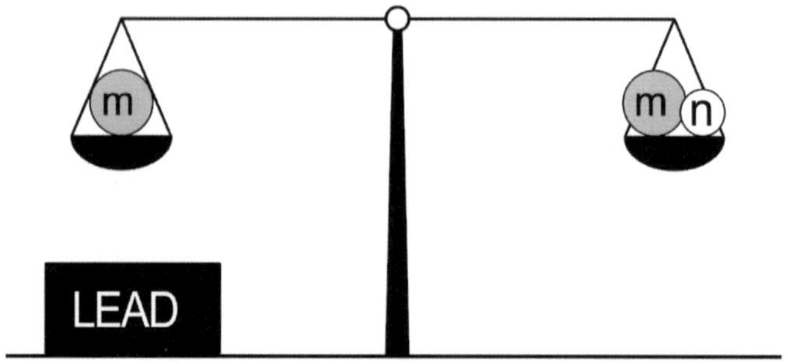

figure 5.13 This event does not take place for Earth and objects attracting each other.

Issac Newton is said to have employed the method of a balance with a chunk of lead to measure the mass of Earth. Putting a chunk of lead under a dish of the balance makes it slightly tip. To get the bar to balance back, an extra mass, which is n in the depict, is loaded in the dish. There forms up a proportion relation between two sides. Thereby the mass of Earth can be extracted. The premise to make this math possible is the universal law of gravity, that is, all the objects attract one another. The lead pulls the mass of the left dish when it is placed under it. This seemed to actually happen and had Issac Newton believe in the universal law of gravity. But the truth is not. The chunk put there means it has to consume some amount of NahL and BCi coming down from between the dish and the lead. So the density of NahL and BCi gets lower than the other side under the other dish. It causes more pressure from above the left dish. That's why it tips a bit to have the lead under it.

As shown in the experiment Newton did to measure how heavy Earth is, all the beings affect one another through space which modern science has neglected and ignored. For this cause old men were always cautious and thoughtful of the environment when they intended to construct something. They did not looked upon resources as their own but as hired for a while, because they thought there is a true lord using and to use them.

An alarming worry, concerning human use of space, namely the satellites in the sky, is there are too many crafts up there and ever getting more. They can affect the environment down here where we spend most of our lives. Or they might be already damaging it. Nowadays concerns are two much traffic, air pollution, plastic garbage, dwindling drinkable water and so on. Why not worry about so many satellites in the sky. It is said there are thousands of satellites already up there. Their total mass might not be so daunting but they can disturb celestial bodies' works on other creatures. No one knows what disaster will and is happening by so much traffic in the sky. It can be, all human members' and all lives', upon the earth, burning or peeling away to disappearance in a second. Plus, overpopulation in the sky may probably be the cause of, more frequent and powerful, recent earthquakes and hurricanes or typhoons, regardless of whether Earth is flat or round.

Appendix

Practices of M and Yang

Reading the writing 'about a language' at the first page of the book would help understand the appendix.

1. Music

A band of people in the costume leads the town. They hold four kinds of percussion instruments and a sort of flute, which is optional. They together with other town people visit each household and play and drink heartily to bless the family. This is what I actually witnessed on a feast day when I was a kid. Now this has become a traditional performance.

The band wear the top black and the bottom white with long ribbons of three colors girdled, blue, yellow, and red. The black and white represents M and Yang, three colors, heaven, earth, and man.

When it comes to M and Yang, many tend to be reminded of Taoism, or the Eastern profound metaphysical philosophy, which sounds very scholarly. M and Yang philosophy has been a daily living culture in Korea. They were mostly farmers or lower class people. Buddhism, Confucianism, and Taoism are the mayor religions(or philosophies) which prevailed in the East Asian society. But they are 2500 years old at oldest. There must have been another one that people in East Asia had believed before those three. M and Yang, heaven, earth, man philosophy precedes them. The specific contents of M and Yang had been missing and dim. Sejong the great revived them through the creation of the alphabet.

2. water

In the beginning God created the heaven and the earth. And the earth was without form, and void; and darkness was upon the face of the deep. And the spirit of God moved upon the face of the waters. (Genesis)

Jesus answered, Verily, verily, say unto thee, Except a man be born of water and of the Spirit, he can not enter into the kingdom of God. (John)

Isn't it weird that there had been water before nothing was created?
It says light was created for the first. Let's see why?

믈(water) 1. water
2. a flock of animals, a multitude of humans
3. (verb) to bite, yet to chew
4. (verb) to pay for something, yet to use it

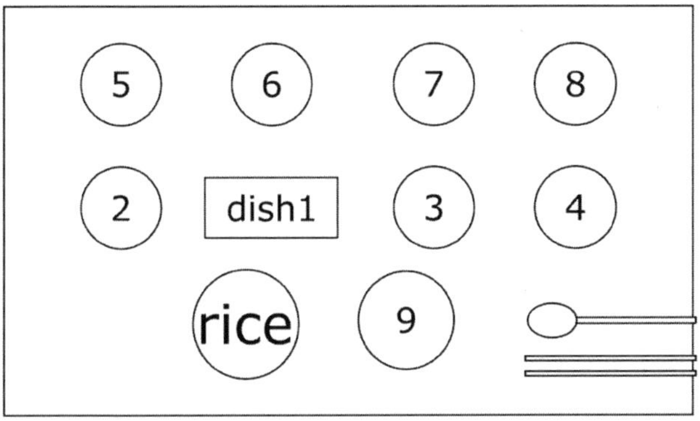

a typical Korean dinning set

The point of the dining table in respect to M and Yang is the somewhat bigger bowl of rice. The number of the other dishes would be more than 20 for a special days, for example, wedding parties. That custom had made me so wonder what made Korean people dine that way. Rice is not a native in Korean peninsula.

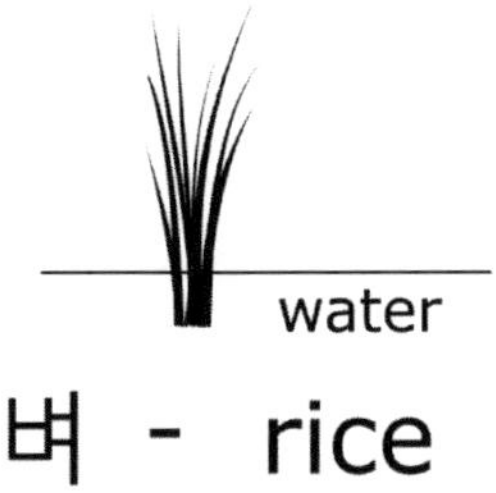

A star and rice plant sound identical in Korean with the letter 'ㄹ' removed from the Korean word meaning star. The letter whose sounding value is between L and R stands for a sin wave. Stars have it. Rice does not. To interpret it, stars and rice do the same job but stars keep doing it periodically. Rice plant does the same for one time, on the other hand. Recall what stars' job was. It was producing NahL. What is rice plant's job? It is producing rice grains, which when cooked, is called BahB.

cotton is a water of NahL and BCi.
Water is a raw material for a production

Why don't you remind yourself where rice plant grows? Right, it carries out its mission in the middle of water. Then it sounds like stars have got to do it in the same environment. Warp and weft threads are from cotton. What about NahL from stars and BCi from the sun? Just like electricity and magnetism are fundamentally the same material, NahL and BCi feel like they have got to share the same origin. In terminology of M and Yang, the origin is called water. Thus, water of warp and weft thread is cotton. A flock of animals is called water in Korean language too. When tamed and used, the wild become an ox, a horse, a dog, and so on.

And seeing the multitudes, he(Jesus) went up into a mountain: and when he was set, his disciples came unto him; And he opened his mouth, and taught them, saying. Blessed are the poor in spirit; for theirs is the kingdom of heaven. (Matthew 5)

That is so widely known the scene of 8 blessings gospel. Doesn't it look like transforming wild animals(multitude-water) into domestic ones(disciples)?

Let's compare absorbing nutritions and maintaining our bodies to embroidering. Base cloth can be NahL and threads of colors BCi. In the Korean dinning, rice plays the role of NahL. It is a base for other dishes. Rice grain is from water as well as NahL from water in the heaven. The other dishes are coloring threads. What and how Koreans eat is a reflection of M and Yang principles. Korea is a kingdom of fermentation foods. The fermenting process is no more than taming beasts, which are also water in the sense of Buddhism as well as M and Yang philosophy.

3. rice and value

Rice is not just a food in Korean culture. A westerner once visited Korea and commented, "Korea is the country of rice straw civilization." It was really so even around the time I was very young. To my memory, there were bags, sacks, mats roofs thatched, of rice straw. The most remarkable thing is they wore shoes woven with rice straw. Rice was fuel for heating, feed for livestocks. I remember the ox my home had rice straw stew for meal. Chaff and bran of rice was a good fertilizer. Ropes were made of rice straw. Rice wine, spirits, many sorts of rice snack. Rice really was NahL of Korean society. And rice was money.

Koreans, mostly my father generation, tended to say, "I am going to sell rice.", which means they had a plan to buy rice. And they said again, "I am going to buy rice at next market.", which means they would have sold rice. I once asked my mother about that when I was an elementary school student. she just smiled at me. She did not know why? They did not speak that way for the other articles, only for rice. It was because rice was used as money for transactions. The transactions by rice are also in my memory.

Buying something with money means selling money by buying the something. There were also metal coins and paper currencies when I witnessed the selling and buying. In respect to rice, such a currency becomes a commodity. This is a little confusing. Suppose my mother bought rice with handing over paper money at the market place. To my mother's viewpoint, she has got money(rice) and released a commodity which is the paper. Getting money for giving up a thing to the other is the definition of selling. This is how the idiom appeared in Korean society.

Rice had long been the measure of value in Korean society. This may look very behind the modern capitalism. It is heavy and bulky as money. But it signifies so much how rice became a yardstick of other values. Rice, as food, is from water and play a role of the base cloth for embroidery in the dining table. It is an independent crystallization of all the participants' labor to produce it. The water, which includes H_2O, carbon dioxide, minerals, etc, is anyway from the heaven through the sun, the moon, and stars. This time the water becomes NahL, light is BCi, and human's labor is the mooning. Rice grain is produced finally. The others are regarded for granted but human labor is not free. So economically Karl Marx was right. What ones are confused about is that technologies create value or at least they help to create. But it is wrong. What so called technologies do is not creating value but increasing capitalists' profit. They help increase the speed of labor exploitation. Value can set in a commodity when consumers are in need for it. Technologies are meaningless without consumers' assessment. At this moment understanding why humans do such things is important. Value is actually permanent but the vessel holding it decays. The value of rice harvested this year may be a half next year and almost naught the year after the next year. What is obvious is value exists to be consumed and transform into more value, and that it should be quick. This tendency of the universe is, in other words, the laws of thermodynamics. But greedy humans want to keep it as long as possible. And they invented interest. In an ideal economy banks keep guests' money to charge them storing fee. The charge is the banks' profit. So the money will get less in the bank as time goes by. Instead, borrowers borrow money from the banks for free. When deposited money is lent, its eating away stops. The borrowers just return it without paying any interest. Then you might say no one would put their money in the bank and the economy does not work. But you know that is how your asset can be preserved for ever in the heaven, said Jesus Christ. I believe this system will actually work, if their faith in the mentioned idea is strong. You can put money in the bank as well as borrow. The effort to preserve or increase an amount of value without further labor is evil. A product secures its value or meaning only when it is used for other value creation. This is a true quantum mechanics of philosophy.

And technologies are also regarded as something descended from heaven in the philosophy of M and Yang. Even though it is admitted that they create value, the idea production is only by labor still has its say. This is a true quantum mechanic of M and Yang. The technology is not determined to be from where until you decided to use how or for what. You use them for good, and they are from heaven. You use them for your own excessive gains, and they are the devil's present. The point is, do not call the devices of an evil intent technologies.

4. weights and measures

1 斤(GN) = 600g

mass of Earth = 5.972×10^{24}kg

1 尺(Zah) = 30cm

velocity of light = 299,792,458m/sec

Modern people's trust in science and technologies has lead to snobbishness and ignorance. However splendid modern technologies are, they are all shabby reflections of real ones on the other side. It can be a coincident but 1 GN is 600g. 1 Zah is 30cm. Those Korean tradition weights and measures were set in pursuit of compatibility to the western metric system. It would have been awkward if 1 GN had been 589g and 1 Zah 29.47cm. But it looks well done because it seems 1 GN and 1 Zah might really have been 600g and 30cm. Historically, they were approximately those numbers. As a modern person, it can not be helped thinking why it has to be 600g, not 500 nor 900. They seem to have known the mass of Earth as well as the velocity of light.

GN is a weight unit. It's got to be significant. Look at what the mass of Earth is. And now what 1 Zah (30cm) signifies? 30cm is 0.3 m, namely 1 / 10000000 of the distance light can go a second. The definition of modern 1m is 1/ 10000000 the shortest distance between the Arctic and the equator. But this is on round Earth, which is wrong. And the surface is rough. Such a calculation was inaccurate from the beginning. Recently its definition was changed to 1 / 10000000 the distance light travels for a second. This was not an official description but it results in that wording in the previous sentence. We need to make sure modern scientific knowledge is not so magnificent as we believe. Those traditional weights and measures are at least 3000 or 4000 years old.

5. man and purchase

Man is Sah-ram in Korean language. Man in the world of M and Yang is not a human being. The bible begins with the creation of man and Korean mythology with a son of God descending from the heaven to multiply men. It is two syllabled word. Each syllable is its own expression of M and Yang. Let's check it out.

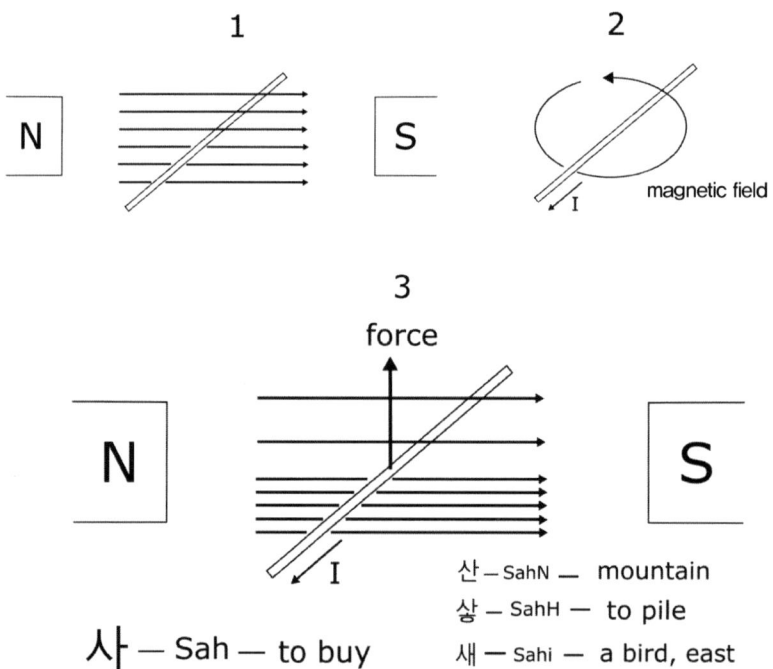

1

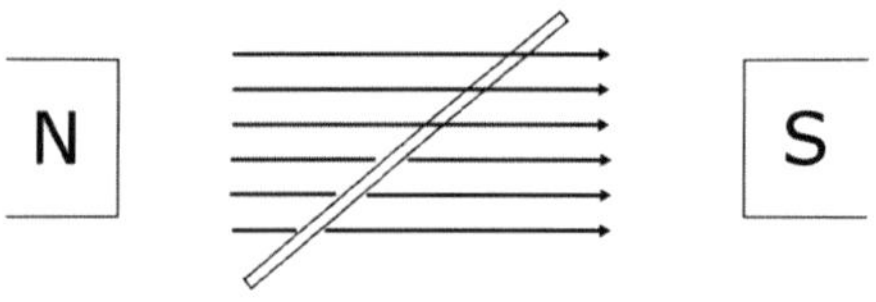

2

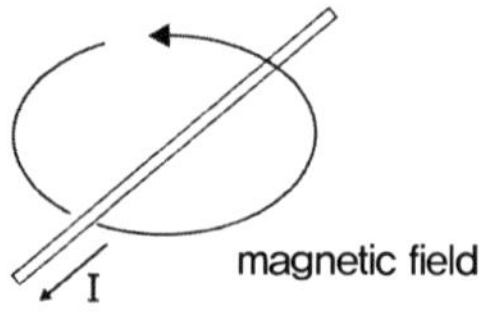

3

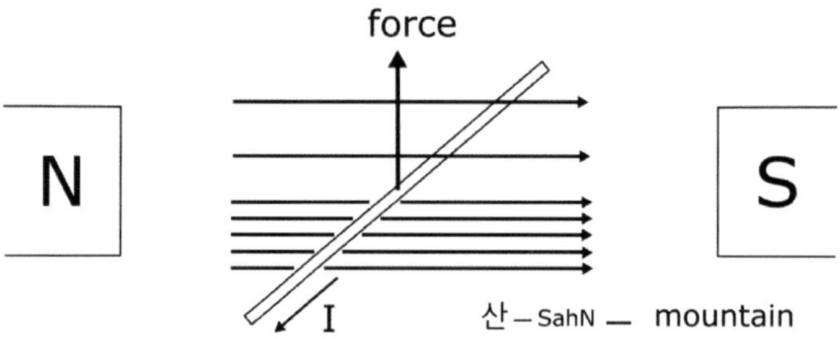

사 — Sah — to buy

산 – SahN — mountain
샇 — SahH — to pile
새 — Sahi — a bird, east

'Sah' stands for the magnetic structure as in the figure. That is a picture of how Fleming's left hand law works practically. The meaning of Sah as a verb is to buy. Let's think of the electric current in the metal bar as money to see how it turns out so. The force generated by Fleming's law would be the commodity you buy to get. The sound N(ㄴ) in Korean language stands for completion. The completed action of Sah is SahN, which means mountain, As seen in the figure, the magnetic flux lifts up and stacks up as the meanings of the derivatives of Sah. The vowel 'i(ㅣ)' symbolizes as man in the Korean alphabet system. So the word Sahi gets to mean a doer of the action 'Sah'. It is a bird in Korean, eventually. This is how all the Korean words are crafted.

買 = 口 + 儿 + 貝

(buy)

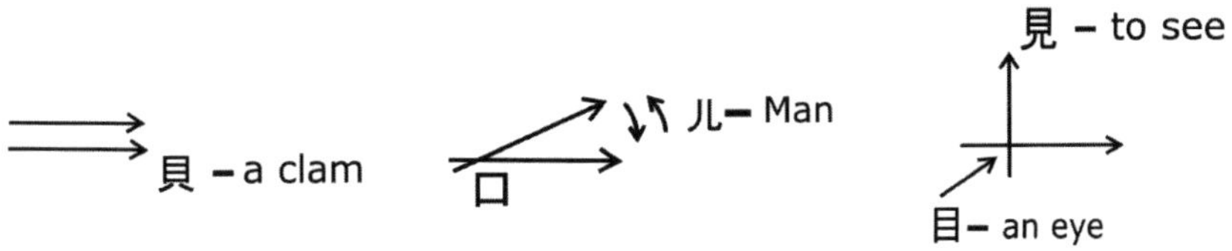

That Chinese letter means to buy. It also turns out to draw the same picture of Sah. Intriguingly, Korean people call the point formed by two lines crossing, an eye. It is because there is light generated by M and Yang crossing. To interpret the letter meaning to buy, having M and Yang crossed at the mouth to fashion the state of clam is the figure of purchase.

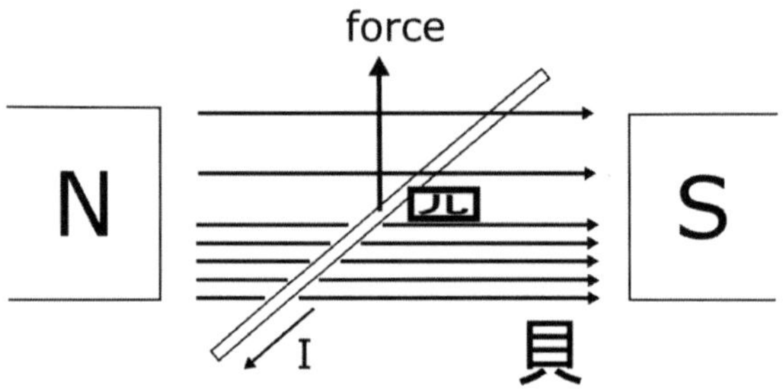

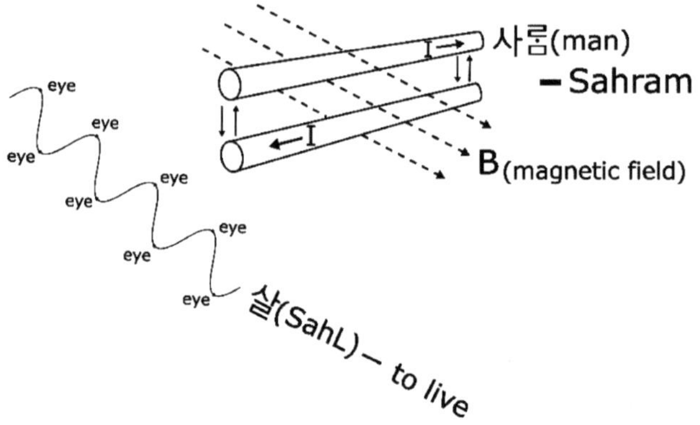

The metal bar would get down, with the direction of the current switched. Keep switching it would lead to a vibration and the altitude of the bar to time should be a sin or cos curve. That action of keeping changing the electric current is Ram. To denote the sin curve in Korean word, it is SahL, which as a verb means to live. The scene is a clear picture of what man is actually. Nevertheless, it feels rather obscure of a legitimate concept of man. That will be coming up in the transistor talk.

6. justice

1

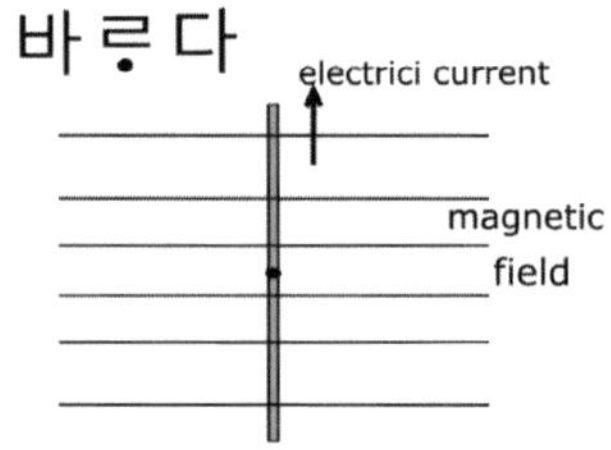

正- 정- JyuhNG

2

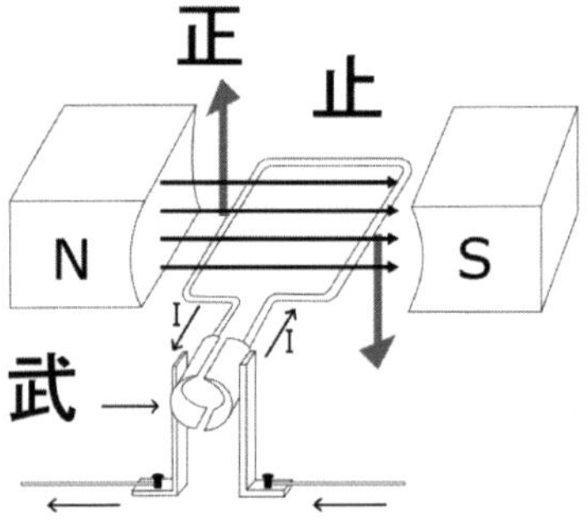

3

stop

正 止(Ji) 正

武 (Moo)

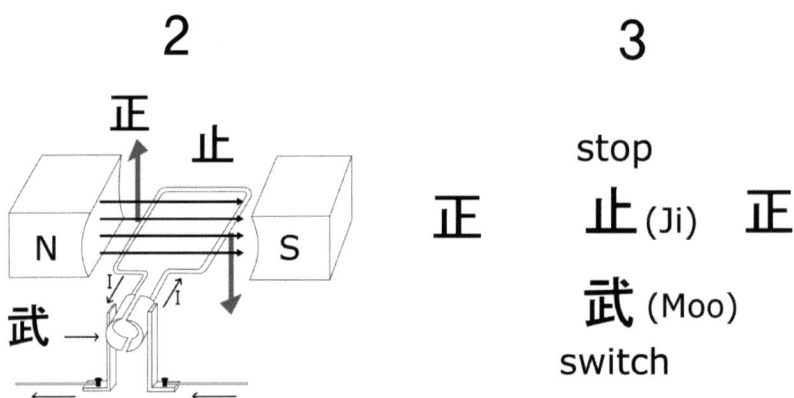

switch

Both an electric motor and an electric generator work in the principle of Fleming's left hand law. A motor generates Kinetic energy with electricity. An electricity generator turns Kinetic energy into electricity in the same principle. In either way, it stands for the Korean word meaning 'right' or 'straight' when two components are right angled to each other. The efficiency at the moment is the highest practically. The Chinese letter features the identical scene and its pronunciation is JyuhNG. Etymologically, its sound seems relatable to English word 'just' and 'justice'. The Chinese letter meaning 'to stop' is a dash missing of JyuhNG and some adding is Moo, which means policing or enforcement. The state that a motor's electric current and magnetic field are at right angle is JyuhNG. And the rotor does not generate any momentum when it erects between the two magnets. So it is Ji, which means stoppage. For the rotor to gain further momentum, it has to change the direction of the electric current. That device is Moo. The letter Moo tends to stand for fighting skills, martial arts, or military force in the Chinese letter using societies. But it is somewhat a misrepresentation. Its accurate definition is an artificial and coercive enforcement toward justice.

7. nation

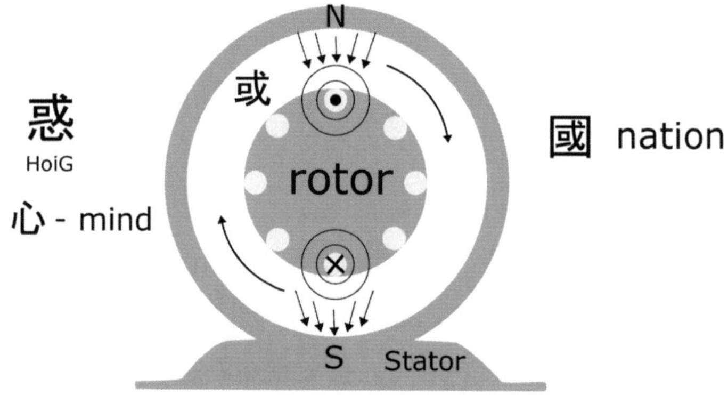

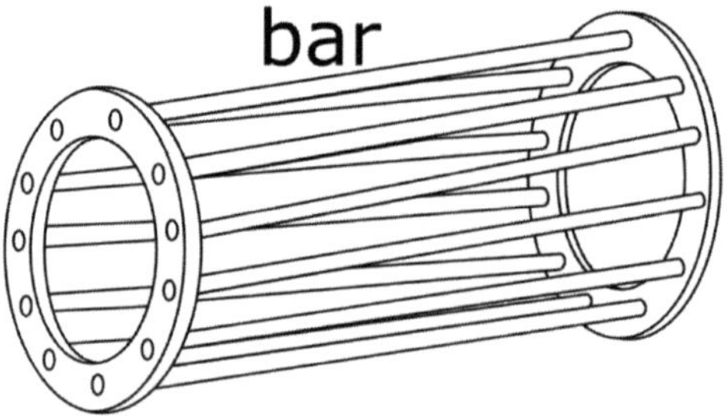

rotor

The induction motor is a very interesting equipment. The classic motor just mentioned is composed of magnets and electric wire. But there is no magnet in an induction motor. The principle is the supplied alternating electric current induces changing magnetic field between the stator and the rotor. The induced magnetic field induces electric current in bars of the rotor. And the induced electric current intensifies magnetic field by a bar of the rotor inducing back in the middle space. Finally the rotor gets pushed to rotate. A Chinese letter to depict this scene of what is happening between the stator and rotor is HoiG, which means illuding. Illuding is also a kind of induction on mind, and a virtual space holding HoiG is a nation. In other words, it's that what a nation is can be something like the inter-inductive systematic operation of an induction motor.

8. man in the virtual(actually real) space

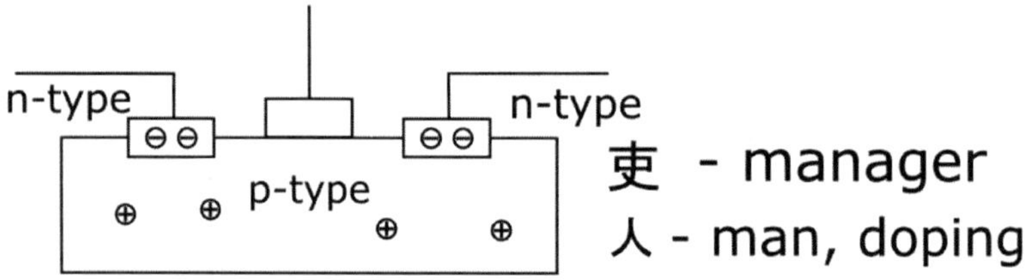

The drawing is a transistor. A nowadays IC chip is a chip where millions or billions of transistors are integrated. Thanks to the invention of this, almost all the modern electric devices and gadgets have come into existence. There are two major roles of a transistor. One is amplification. It amplifies input signals by 50 times or 100 times. This is how a singer can sing in the huge stadium.

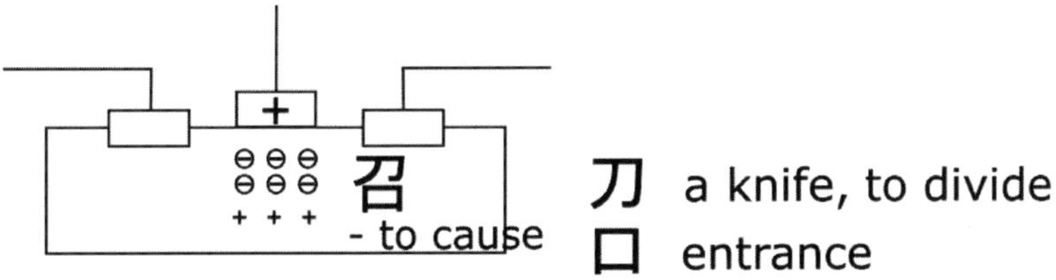

The other is it controls an electric current so that it can produce a binary signal, 0 and 1. Here we can see M and Yang inside it. The gate's polarization to the plus divides the area between the n-type semiconductors into the positive and the negative band, and the electric current flows.

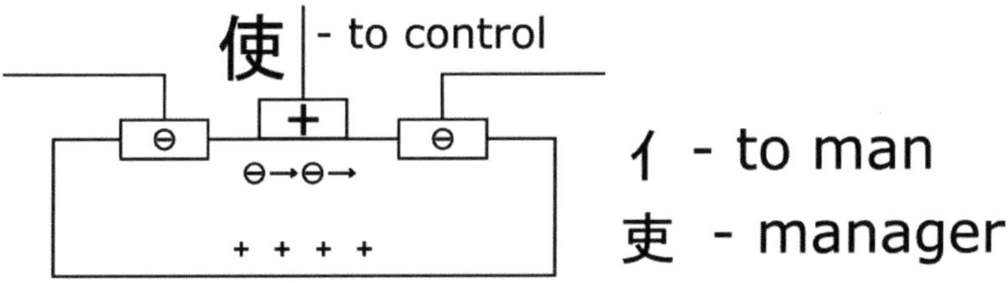

To interpret what man is through the Chinese letter and the pronunciation by the publication of Sejong the great, man is a controller in the world of M and Yang. What we experience are their performances.

人 신

9. superposition

It is said a quantum computer is around the corner and this computer's capability is beyond our imagination. The principle it works on is superposition. A classical computer operates on a binary bit. It is composed of 0, 1. But a quantum computer's bit is somewhat different. The point of how this computer works is it uses the state of ' between 0 and 1'. For example, an electron has two spins. It can be used as a bit. Up spin is 0 and down 1. What about the middle state? In the middle state it can be down as well as up at the same time. But once confirmed, it wakes out of the state. Such a state does not exist in reality. Our recognition of a being in such a state breaks the state away. So researchers try to lower the temperature extremely in order to have the operation environment superposition friedly which is really hard to get to. But there is a place where the state of superposition can be easily built. It is where there is a mind. The M and Yang founders knew beings of such states. The state of superposition we learned at the physics class is not a true superposition because it resulted from inability of our sensory organs and lack of their subtlety. That of the double slit experiment means we do not know which slit an electron has been through. It was not that an electron passes through two slits at the same time. Anton

Zeilinger' is the same case. Just we do not know how it happened. All the cases of superposition issued by modern physics are actually fake ones. All are caused by our ignorance. However, that does not mean there is no such one in the world.

3 4 3 4

3 4 3 4

3 5 4 3

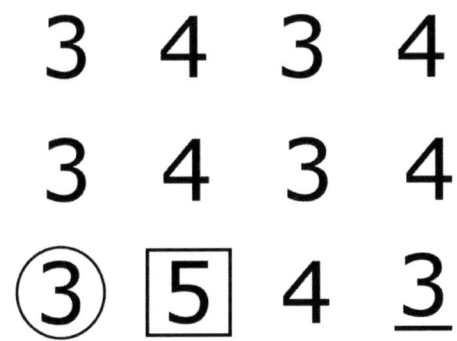

Modern scientists and engineers have just started to study this area. But the founders knew and made use of it. The products are music, art, literature, and so on. The lineup of numbers above is the structure of an old Korean fix verse. Each number marks the number of syllables in the phrase. The point of the structure is number three. Korean culture seemed to me, obsessed with number three. The number of rows of the Korean old poems is three. The most important phrase of the poem is the first of the third row. It is also three. And the second important phrase is the second one of the third row, which is five. They are all odd numbers. That literature changed into a free style as time went by. But the number of the first of the third row remained unchangeable and untouchable.

The purpose of music is touching and moving. They come about when our minds are put in a superposition state. And this is also why number three is so important in Korean culture. 2 and 3 are the representatives of all natural numbers. 2, of even numbers and 3, of odd numbers. When it comes to superposition, what comes up on your mind? If my feeling did not fail to get in common with others', it would be 'ambiguity'. On the other hands, what is the feel of an even number, two or four? It is finished, determined, stable, and so on. An odd numbers are unfinished, unstable, undetermined and so on. The first and second phrase of the third row is the climax of the work. They are odd numbers. And the last one is also three syllables. The effect the composer intends with the last as an odd number three is incompleteness. The ordinary Korean sentences end with 'Dah'. Sejong the great, who invented the Korean alphabetic system, also wrote a literary work. All the sentences in his work end with 'Ni'. It gives someone like me, who is a native speaker, the feeling of "Hey sir, you want to say some more." All these literary but mathematical measures are relevant to quantum mechanics of mind.

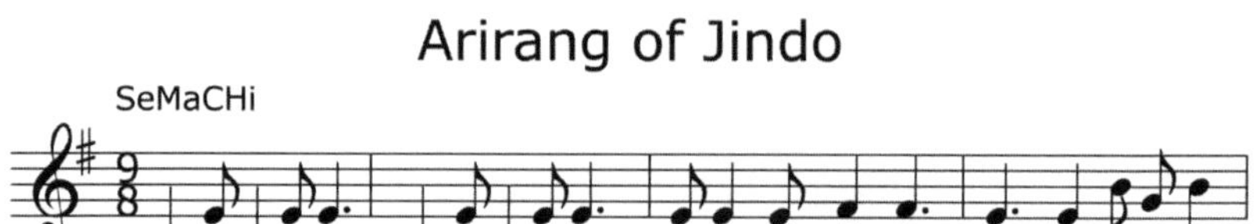

As long as I remember, all the Korean traditional folk songs but one are triple time, while modern musics I listen to recently are binary time. They are mostly four-four time. But it is a repetition of binary time. Korean traditional ones are six-eight, nine-eight, and even twelve-eight time, which are times of triple. The percussion musics played by the band remarked at the first corner are composed of varieties of three beats and two beats. They are all folk musics, not like the ones by Mozart and Beethoven.

Those three beats music are often used to call out ghosts. In descent words, they were played in a religious rite. Though it is the last moment of the paper, the kernel of study of M and Yang lies in what the spiritual are, to emphasize the essential use of M and Yang.

The study has been doing in divided areas, humanities and natural science. It was not so at the beginning. What matter is is an illusion, after all. There must be something producing illusions. This is why studies dividing into humanities and natural science, or science verses religion, physics verses metaphysics, is absurd.

Publication Garlic and Onion

Publication Garlic and Onion aims at the propagation of Tai-gg (M and Yang) philosophy. A bulb of garlic is composed of sections, while that of onion layers. They represent M and Yang, and the East and the West, as a garlic is a native in Korea and an onion was imported to grow.

- **address** : 191 Gujidongro Guji Dalsung Daegu Republic of Korea
- **e-mail** : hotonion153@hanmail.net
- **phone** : 010-7242-5315

First published January 2020
Designed and printed by publication Bookaa, South Korea

ISBN 978-89-964082-4-6

Keywords for would be uploaded Youtube clips ;
M and Yang, physics of M and Yang, Korean language